AF681547

ENERGY POLICIES, POLITICS AND PRICES

UTILITY-SCALE SOLAR INSTALLATIONS

BACKGROUND, POLICIES, AND FINANCIAL STRUCTURES

ENERGY POLICIES, POLITICS AND PRICES

Additional books in this series can be found on Nova's website under the Series tab.

Additional e-books in this series can be found on Nova's website under the e-book tab.

ENERGY POLICIES, POLITICS AND PRICES

UTILITY-SCALE SOLAR INSTALLATIONS

BACKGROUND, POLICIES, AND FINANCIAL STRUCTURES

MILES B. SIMMONS
EDITOR

nova publishers
New York

Copyright © 2014 by Nova Science Publishers, Inc.

All rights reserved. No part of this book may be reproduced, stored in a retrieval system or transmitted in any form or by any means: electronic, electrostatic, magnetic, tape, mechanical photocopying, recording or otherwise without the written permission of the Publisher.

For permission to use material from this book please contact us:
Telephone 631-231-7269; Fax 631-231-8175
Web Site: http://www.novapublishers.com

NOTICE TO THE READER

The Publisher has taken reasonable care in the preparation of this book, but makes no expressed or implied warranty of any kind and assumes no responsibility for any errors or omissions. No liability is assumed for incidental or consequential damages in connection with or arising out of information contained in this book. The Publisher shall not be liable for any special, consequential, or exemplary damages resulting, in whole or in part, from the readers' use of, or reliance upon, this material. Any parts of this book based on government reports are so indicated and copyright is claimed for those parts to the extent applicable to compilations of such works.

Independent verification should be sought for any data, advice or recommendations contained in this book. In addition, no responsibility is assumed by the publisher for any injury and/or damage to persons or property arising from any methods, products, instructions, ideas or otherwise contained in this publication.

This publication is designed to provide accurate and authoritative information with regard to the subject matter covered herein. It is sold with the clear understanding that the Publisher is not engaged in rendering legal or any other professional services. If legal or any other expert assistance is required, the services of a competent person should be sought. FROM A DECLARATION OF PARTICIPANTS JOINTLY ADOPTED BY A COMMITTEE OF THE AMERICAN BAR ASSOCIATION AND A COMMITTEE OF PUBLISHERS.

Additional color graphics may be available in the e-book version of this book.

Library of Congress Cataloging-in-Publication Data

ISBN: 978-1-62948-999-5

Published by Nova Science Publishers, Inc. † New York

CONTENTS

PREFACE

Solar energy technologies continue to be deployed at unprecedented levels, aided significantly by the advent of large-scale projects that sell their power directly to electric utilities. Such utility-scale systems can deploy solar technologies far faster than traditional "behind-the-meter" projects designed to offset retail load. These systems achieve significant economies of scale during construction and operation, and in attracting financial capital, which can in turn reduce the delivered cost of power. Topics discussed in this compilation include a technology and market overview of utility-scale concentrating solar power and photovoltaics projects; federal and state structures to support financing utility-scale solar projects and the business models designed to utilize them; and the impact of financial structure on the cost of solar energy.

Chapter 1 - This is the first in a series of three reports on utility-scale solar installation in the United States. This report serves as: (1) a primer on utility-scale solar technologies and (2) a summary of the current state of the U.S. utility-scale solar market. The second report overviews policies and financing of utility-scale solar systems; the third report assesses the impact of financial structures on the cost of energy from utility-scale systems.

Utility-scale solar projects are generally categorized in one of two basic groups: concentrating solar power (CSP) and photovoltaic (PV). CSP systems generally include four commercially available technologies: CSP trough, CSP tower, parabolic dish, and linear Fresnel reflector, although only CSP trough and CSP tower projects are currently being deployed. CSP systems can also be categorized as hybrid systems, which combine a solar-based system and a fossil fuel energy system to produce electricity or steam.

PV systems usually include either crystalline silicon (c-Si) or thin-film technologies. Thin film includes an array of advanced materials, but only one—cadmium telluride (CdTe)—has had significant success in utility-scale solar development. Additionally, this report covers concentrating photovoltaic (CPV) systems, which only recently have gained traction in the utility-scale market with several signed contracts.

According to a database maintained by the National Renewable Energy Laboratory (NREL), there are approximately 16,043 megawatts (MW) of utility-scale solar resources under development in the United States as of January 2012. PV projects make up the overwhelming majority (about 72%) of facilities under development. While many developers have specified that their projects will use PV (e.g., c-Si or CdTe), in some cases the technology will be selected just prior to construction. This selection will likely depend on module pricing at the time of order placement once all necessary permits have been obtained and pre-construction activities completed. It is not uncommon, especially given the recent drop in c-Si module prices, for developers to switch technologies in the planning phase.

According to NREL's internal database, CdTe thin-film technology represents about one-fifth of the total inventory of planned utility-scale solar projects and nearly one-third of total planned PV projects. First Solar was once considered the sole or joint developer of all utility-scale CdTe projects under development in the United States, though this is changing with the entrance of General Electric into the CdTe market. GE is currently contracted to supply panels to the 20 MW Illinois Solar plant being developed by Invenergy.

Approximately 8,224 MW of developing projects are utilizing c-Si modules or have not indicated final technology selection. The majority of these projects are expected to select c-Si- based modules. Per NREL's criteria—5 MW or larger and holding a long-term contract— approximately 11,500 MW of total PV capacity is under development in the United States, including c-Si, CdTe, copper-indium-selenide (CIS) modules, and CPV technologies.

Among CSP projects, tower systems have a slight market penetration edge over parabolic troughs (about 16% versus 9% of all utility-scale solar systems under development). NREL's project announcements database indicates that the tower market is dominated by one developer, BrightSource Energy, who holds over 2.2 GW of PPAs with California utilities. Solar Millennium was the principal developer in the trough space, but the company's announced switch to PV and subsequent sale of all proposed projects to solarhybrid has left only six trough developers and no clear market front-runners.

Two cutting-edge solar technologies, Enviromission's solar chimney and Solaren's space solar project, are indicated as "Other" because they hold PPAs and constitute significant additions to the total capacity under development, but they are not categorized as traditional CSP or PV technologies. Two solar/fossil hybrid plants representing a combined 100 MW of solar capacity are included as a separate category to note their distinct approach; both plants will use solar power to supplement natural gas-fired generation. Finally, CIS is included because of the recent announcement that Solar Frontier, the Japanese CIS manufacturer, will supply up to 150 MW of panels to energy developer enXco for use in their PPA contracts with San Diego Gas and Electric (SDG&E).

Currently, multiple utility-scale systems are producing power on a consistent basis. The nine solar trough CSP plants that comprise the solar energy generating system (SEGS) in California's Mojave Desert constitute the majority of CSP. The SEGS units commenced commercial operation from 1984–1991 with several additional utility-scale CSP projects coming online recently. In 2007, the 64 MW Nevada Solar One project, a CSP trough plant developed by Acciona Solar Power, became operational. Two 5-MW demonstration facilities developed by Ausra and eSolar also became operational in 2008 and 2009, respectively. There are over 40 utility-scale PV facilities currently operational in the United States, amounting to some 673 MW of capacity.

Chapter 2 - Over the past two decades, state and federal policymakers have enacted numerous policies to support new renewable energy development. The following report evaluates those policies based on industry literature, publicly available data, and questionnaires conducted by the National Renewable Energy Laboratory (NREL). As such, the report is relevant to policymakers who want to enable project financing via effective legislation, to regulators interested in supporting new project development, and to new investors interested in entering the renewable energy project finance space.

Three policy types have been critical to the widespread deployment of utility-scale solar facilities:

1. Federal tax benefits comprising the 30% investment tax credit (ITC) and accelerated depreciation schedules, and the Section 1603 cash grant program available in lieu of the ITC
2. The Department of Energy (DOE) loan guarantee program, which has provided guarantees and access to low-cost financing, enabling larger solar energy projects than ever deployed before

3. State renewable portfolio standards (RPS), enacted by 29 states and the District of Columbia (D.C.), which have created a necessary demand for renewable energy, and in particular, have led to the signing of long-term contracts critical to financing large-scale solar facilities.

As of February 21, 2012, the Section 1603 (also known as the "cash grant") program had awarded $10.76 billion in grants to roughly 23,000 projects. The cash grant program expired in 2011 but will award projects that started construction in 2009, 2010, or 2011.

Utility-scale solar projects in particular benefitted significantly from the loan guarantee program. The loan guarantee program provides access to capital at a lower cost and larger quantity than what private markets would provide without the program. Through January 2012, solar projects have been supported solely under Section 1705 of the loan guarantee program, through which over $12 billion in loans have been made to 3,500 MW of solar projects. To support the solar project loans, an estimated $1.4 billion in credit subsidy costs have been paid under the American Recovery and Reinvestment Act.

The credit subsidy is essentially a fund set aside by the DOE to cover the costs associated with specific project failures. As the loan guarantee program required an average credit subsidy of 11.7% per project, every dollar the DOE spends in credit subsidy will support $8.55 in loans. Assuming an average financial structure of 60% debt and 40% equity (actual equity investments are not known), each credit subsidy dollar invested can actually be extended to support $14.25 in solar project development.

Of note, the private bond market—without any loan guarantees—has rebounded from the financial crisis. In February 2012, to finance the Topaz purchase, MidAmerican Energy successfully offered $850 million of private placement bonds and is considering a second bond offering to raise an additional $430 million.

State RPSs create critical demand for large-scale generation projects, driving investor- owned and public utilities to sign long-term purchase power agreement (PPA) contracts with solar project developers. These PPAs are fundamental to project financing as debt and equity investors will not risk capital without the surety of revenue from a creditworthy counter-party to the contract.

However, the solar industry is capable of far greater deployment than currently required under the collective RPSs presently enacted. The Lawrence

Berkeley National Laboratory recently estimated that nine states and D.C. have specific solar renewable energy certificate requirements and will require incremental solar capacity of 6,729 MW by 2025. That represents roughly 480 MW of solar capacity installed per year, less than half the total domestic installations in 2010. Even in the most aggressive state—New Jersey—current solar project deployment meets the build-out requirements for 2025.

Although the array of federal and state support mechanisms greatly improved the ability of solar energy projects to achieve financing, complex financial structures were still required to utilize the depreciation benefits and access investment capital. The advanced financial structures described in this report include:

- Single owner finance, which includes utility balance sheet financing and developer balance sheet financing
- Lease structures generally referred to as sale-leaseback and inverted leases
- Partnership flips
- Utility prepay (which has been solely utilized by the wind industry only modestly to date).

This report is NREL's second study in a three-part series on utility-scale solar in the United States. In the first study, NREL found that, as of January 2012, approximately 16,000 MW of utility-scale solar projects (defined as 5 MW or larger) are in various stages of advanced development and hold a contract with an investor-owned or public utility. If all these projects come to fruition, the amount of solar in the United States will increase approximately six-fold. The third study is a quantitative analysis on the cost impact of specific financial structures and terms as they relate to utility-scale solar facilities.

Chapter 3 - To stimulate investment in renewable energy generation projects, the federal government developed a series of support structures that reduce taxes for eligible investors—the investment tax credit, the production tax credit, and accelerated depreciation. The nature of these tax incentives often requires an outside investor and a complex financial arrangement to allocate risk and reward among the parties. These financial arrangements are generally categorized as "advanced financial structures." Among renewable energy technologies, advanced financial structures were first widely deployed by the wind industry and are now being explored by the solar industry to support significant scale up in project development. This report describes four of the most prevalent financial structures used by the renewable sector and evaluates the impact of financial structure on energy costs for utility-scale solar projects that use photovoltaic and concentrating solar power technologies.

A critical aspect of the analysis is the input assumptions, including the cost to install each technology, the costs and terms of financial capital relevant to the technology and financial structure assessed, and the operating costs of each technology. To determine reasonable inputs, the authors conducted a series of interviews with industry experts, reviewed several sources in the current literature, and relied on the default values from the System Advisor Model that was employed in this analysis, which were, in turn, guided by the industry expertise of its developers.

The analysis determined that financial structures that include project-level debt generally yield a lower levelized cost of energy (LCOE) compared to those that rely purely on equity capital, although in practice raising debt at the project level can be difficult, particularly for developers without sizable balance sheets and a strong history of development experience. Other insights from the analysis include:

- Debt associated with the loan guarantee program can reduce LCOE by approximately 20%, and possibly more, depending on the quantity of debt the project is allowed to take on.
- Certain financial experts expect "tax equity" capital—incorporated to maximize the use of federal tax credits and depreciation benefits—to increase in cost by 200 to 400 basis points (bp) (or 2%–4%) due to the termination of the current 1603 Treasury grant program at the end of 2011. Such an increase in the cost of tax equity is projected to raise the LCOE from utility-scale solar projects by 3%–20%.
- The cost of tax equity is particularly critical in all-equity financial structures and can have a significant impact on the cost of energy produced. Even more important in all- equity financial structures, the internal rate of return (IRR) target year—the project year in which the tax equity investor is expected to earn the pre-negotiated return— can have a very large impact on the resulting cost of energy. Delaying the IRR target year, for example, from year eight of the project to year nine, can improve the LCOE by 7%–27%, depending on the technology and financial structure employed.

- Development and operational experience is projected to lead to further financial structure innovation and reduction in required investment returns for various sources of financial capital.

In: Utility-Scale Solar Installations
Editor: Miles B. Simmons

ISBN: 978-1-62948-999-5
© 2014 Nova Science Publishers, Inc.

Chapter 1

UTILITY-SCALE CONCENTRATING SOLAR POWER AND PHOTOVOLTAICS PROJECTS: A TECHNOLOGY AND MARKET OVERVIEW*

Michael Mendelsohn, Travis Lowder and Brendan Canavan

LIST OF ACRONYMS

AC	alternating current
a-Si	amorphous silicon
b	bar
Bell	Bell Solar Thermal
BLM	Bureau of Land Management
CEC	California Energy Commission
CdTe	cadmium telluride
CIGS	copper indium gallium selenide
CPV	concentrating photovoltaic
c-Si	crystalline silicon
CSP	concentrating solar power
DC	direct current
DOE	Department of Energy

* This is an edited, reformatted and augmented version of the U S. Department of Energy, Office of Energy Efficiency & Renewable Energy, NREL/TP-6A20-51137, dated April, 2012.

FPL	Florida Power and Light
ft	foot
HTF	heat transfer fluid
ISO	independent system operator
IOU	investor-owned utility
MW	megawatt
NERC	North American Energy Reliability Corporation
PEIS	Programmatic Environmental Impact Statement
PG&E	Pacific Gas and Electric
PPA	power purchase agreement
PV	photovoltaic
RPS	renewable portfolio standard
SCE	Southern California Edison
SDG&E	San Diego Gas and Electric
SEGS	Solar Energy Generating System
SGIP	small generator interconnection process
TEP	Tucson Electric Power
TES	thermal energy system

EXECUTIVE SUMMARY

Solar energy technologies continue to be deployed at unprecedented levels, aided significantly by the advent of large-scale projects that sell their power directly to electric utilities. Such utility- scale systems can deploy solar technologies far faster than traditional "behind-the-meter" projects designed to offset retail load. These systems achieve significant economies of scale during construction and operation, and in attracting financial capital, which can in turn reduce the delivered cost of power.

This is the first in a series of three reports on utility-scale solar installation in the United States. This report serves as: (1) a primer on utility-scale solar technologies and (2) a summary of the current state of the U.S. utility-scale solar market. The second report overviews policies and financing of utility-scale solar systems; the third report assesses the impact of financial structures on the cost of energy from utility-scale systems.

Utility-scale solar projects are generally categorized in one of two basic groups: concentrating solar power (CSP) and photovoltaic (PV). CSP systems generally include four commercially available technologies: CSP trough, CSP tower, parabolic dish, and linear Fresnel reflector, although only CSP trough and CSP tower projects are currently being deployed. CSP systems can also be categorized as hybrid systems, which combine a solar-based system and a fossil fuel energy system to produce electricity or steam.

PV systems usually include either crystalline silicon (c-Si) or thin-film technologies. Thin film includes an array of advanced materials, but only one—cadmium telluride (CdTe)—has had significant success in utility-scale solar development. Additionally, this report covers concentrating photovoltaic (CPV) systems,[1] which only recently have gained traction in the utility-scale market with several signed contracts.

According to a database maintained by the National Renewable Energy Laboratory (NREL),[2] there are approximately 16,043 megawatts (MW) of utility-scale solar resources under development[3] in the United States as of January 2012 (see Figure ES-1). PV projects make up the overwhelming majority (about 72%) of facilities under development. While many developers have specified that their projects will use PV (e.g., c-Si or CdTe), in some cases the technology will be selected just prior to construction. This selection will likely depend on module pricing at the time of order placement once all necessary permits have been obtained and pre-construction activities completed. It is not uncommon, especially given the recent drop in c-Si module prices, for developers to switch technologies in the planning phase.

According to NREL's internal database, CdTe thin-film technology represents about one-fifth of the total inventory of planned utility-scale solar projects and nearly one-third of total planned PV projects[4]. First Solar was once considered the sole or joint developer of all utility-scale CdTe projects under development in the United States, though this is changing with the entrance of General Electric into the CdTe market. GE is currently contracted to supply panels to the 20 MW Illinois Solar plant being developed by Invenergy.

Approximately 8,224 MW of developing projects are utilizing c-Si modules or have not indicated final technology selection. The majority of these projects are expected to select c-Si- based modules. Per NREL's criteria—5 MW or larger and holding a long-term contract— approximately 11,500 MW of total PV capacity is under development in the United States, including c-Si, CdTe, copper-indium-selenide (CIS)[5] modules, and CPV technologies.

Among CSP projects, tower systems have a slight market penetration edge over parabolic troughs (about 16% versus 9% of all utility-scale solar systems under development). NREL's project announcements database indicates that the tower market is dominated by one developer, BrightSource Energy, who holds over 2.2 GW of PPAs with California utilities. Solar Millennium was the principal developer in the trough space, but the company's announced switch to PV and subsequent sale of all proposed projects[6] to solarhybrid has left only six trough developers and no clear market front-runners.

Figure ES-1 provides an overview of the U.S. utility-scale solar market. Two cutting-edge solar technologies, Enviromission's solar chimney and Solaren's space solar project, are indicated as "Other" because they hold PPAs and constitute significant additions to the total capacity under development, but they are not categorized as traditional CSP or PV technologies. Two solar/fossil hybrid plants representing a combined 100 MW of solar capacity are included as a separate category to note their distinct approach; both plants will use solar power to supplement natural gas-fired generation. Finally, CIS is included because of the recent announcement that Solar Frontier, the Japanese CIS manufacturer, will supply up to 150 MW of panels to energy developer enXco for use in their PPA contracts with San Diego Gas and Electric (SDG&E).

Currently, multiple utility-scale systems are producing power on a consistent basis. The nine solar trough CSP plants that comprise the solar energy generating system (SEGS) in California's Mojave Desert constitute the majority of CSP. The SEGS units commenced commercial operation from 1984–1991 with several additional utility-scale CSP projects coming online recently (EIA 2008). In 2007, the 64 MW Nevada Solar One project, a CSP trough plant developed by Acciona Solar Power, became operational (Acciona 2010). Two 5-MW demonstration facilities developed by Ausra and eSolar also became operational in 2008 and 2009, respectively (Ausra 2008; eSolar 2009). There are over 40 utility-scale PV facilities currently operational in the United States, amounting to some 673 MW of capacity. See Appendix Table A-1 for a full list of operating utility-scale PV plants.

1. INTRODUCTION

Drivers ranging from energy security and cleaner air to global economic competitiveness and rapidly falling costs are sparking a significant shift in energy generation policy and planning. Electric utilities in the United States and the regulatory agencies that oversee them are increasing renewable energy use to meet electric load. Technological advances in materials and components and heightened experience among market entities are leading the way to more cost-effective renewable power production. Renewables have also significantly benefitted from a raft of support policies and incentives at the municipal, state, and federal levels. These include federal tax credits, cash grants, loan guarantee programs, feed-in tariffs, and state renewable portfolio standards (RPS),[7] which are discussed in detail in the second utility-scale solar report. For example, California's RPS, the most robust in the United States, with a required 33% of renewable generation from its investor-owned utilities (IOUs), has touched off a spate of solar procurement in the last two years. Today, California's three IOUs hold PPAs with nearly 72% of the total solar capacity under development in the United States (see Appendix C).

Supportive policies, financial innovations, and plummeting technology costs have spurred utility- scale[8] solar market development in the United States. This report introduces that growing market. It has two objectives: (1) to summarize solar technologies deployed at utility-scale installations, and (2) to provide a market overview of U.S. deployment activities. The report is divided by technology type: Section 2 deals with CSP technologies, and Section 3 deals with PV solar power technologies. Market overviews for each technology are provided at the conclusion of each subsection. This report only considers projects already contracted to sell power [typically in the form of a power purchase agreement (PPA)].

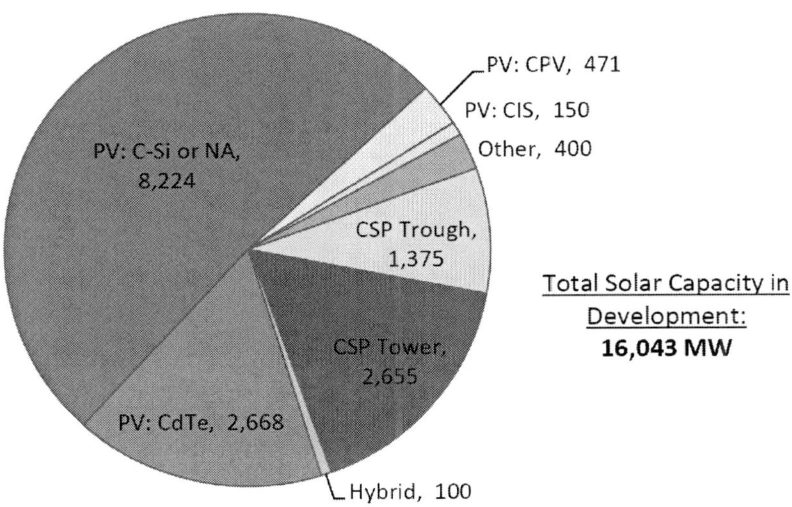

Figure ES-1. Total U.S. utility-scale solar capacity under development (all numbers in MW).

1.1. Utility-Scale Market Overview

Approximately 1,176 MW of utility-scale solar power was operational as of January 2012 (see Figure 1). About 43% (503 MW) of this capacity is furnished by CSP facilities, all but 10 MW of which utilize trough technology; the remaining 57% of this capacity comes from PV installations. Crystalline silicon (C-Si) and cadmium telluride (CdTe) comprise the majority of technologies deployed at these installations with 58.0% and 34.5% representation, respectively. Amorphous silicon (a-Si), another thin-film technology, represents about 7.0% of total PV installations, and concentrating photovoltaic (CPV) about 0.5%.

Figure 2 illustrates that PV capacity will continue to outpace CSP in the United States as more developing projects come online. Nearly all utility-scale CSP plants today use troughs; however, most planned CSP capacity will not use troughs. Instead, CSP towers have become the preferred technology, with over 2,655 MW of projects under contract. CSP tower developer BrightSource holds the majority of the PPAs, with about 2.2 GW of capacity (82% of the total planned CSP capacity). Recent CSP trough market contraction was largely the result of developer Solar Millennium's technology swap for their Blythe, Amargosa, and Palen facilities. At least 2 GW of CSP troughs were scrapped for PV because of what Solar Millennium described as more "favorable conditions in the PV and commercial bank markets" (PV Magazine 2011).

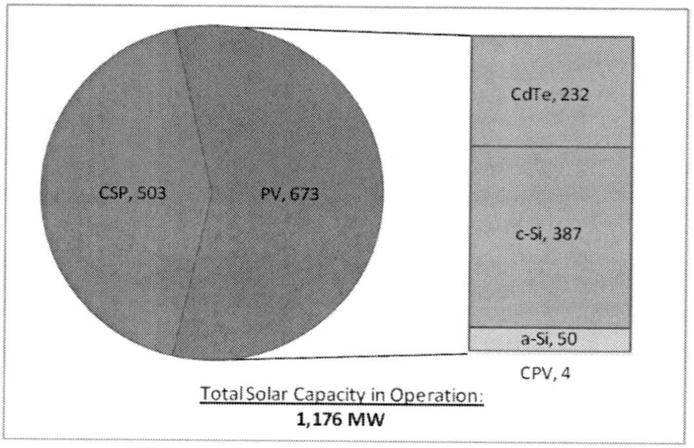

Figure 1. Total U.S. utility-scale solar capacity in operation as of January 2012 (all numbers in MW).

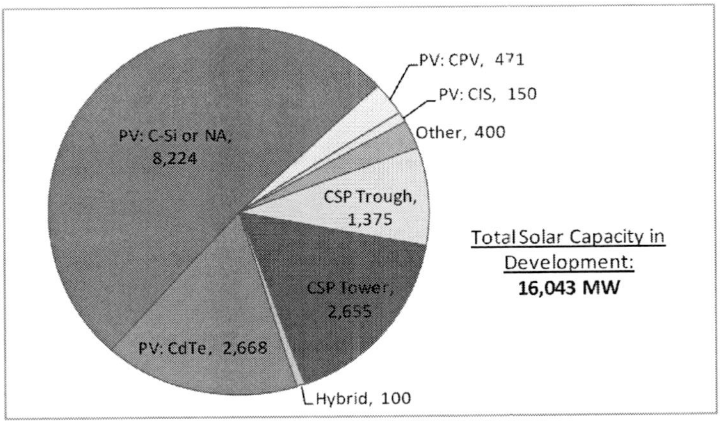

Figure 2. Total U.S. utility-scale solar capacity under development (all numbers in MW).

2. Concentrating Solar Power

CSP systems produce electricity by focusing sunlight to heat a fluid. The fluid then boils water to create steam that spins a conventional turbine and generates electricity or it powers an engine that produces electricity (Richter et al. 2009). CSP plants consist of three major subsystems: one that collects solar energy and converts it to thermal energy; a second that converts the thermal energy to electricity; and a third that stores thermal energy collected from the solar field and subsequently dispatches the energy to the power block.

There are currently 503 MW of utility-scale CSP facilities operating domestically.

The first large-scale, commercial CSP stations were the solar energy generating systems (SEGS) built by Luz International, Ltd. from 1984–1991 (DOE 2010c). Nine plants were built in three separate locations for a total of 354 MW. Figure 3 shows SEGS 4, located in Kramer Junction, California, which has a peak output of 150 MW. SEGS 1 and 2 have a combined maximum output of 44 MW and are located in Daggett, California. SEGS 8 and 9 have a combined maximum output of 160 MW and are located in Harper Lake, California. NextEra operates and partially owns SEGS 3–9, with a combined maximum output of 310 MW (NextEra 2010).

The latest CSP plant to be developed was the 75 MW Martin Next Generation Solar Energy Center developed by and for NextEra subsidiary

Florida Power and Light (FPL). The plant was completed in 2010 (FPL 2010). This facility uses CSP trough technology to supplement the 3,705 MW gas- and oil-fired Martin Generation facility and is considered in this report to be a solar/fossil hybrid plant.

CSP systems are unique in the renewable energy sector in that they can integrate large-scale thermal energy storage (TES). The first utility-scale plants with storage are now operating in Spain (Andasol 1–3) and were developed by Solar Millennium (Solar Millennium 2010). At least six plants with TES are currently in development in the United States—the 250 MW Solana Solar plant by Abengoa Solar (6 hours of dispatchable storage), the 110 MW Crescent Dunes plant by Solar Reserve (10 hours of dispatchable storage), the 5 MW Bell Solar Thermal by Bell Energy (storage capacity unknown), and three BrightSource projects whose locations and storage capacities are yet undisclosed (Wesoff 2010; Wesoff 2011; Environmental Leader 2010; BrightSource Energy 2011a). Solana and Crescent Dunes finalized loan guarantees from the U.S. Department of Energy (DOE) for $1.45 billion and $737 million, respectively, to support project development (DOE 2011c).

Table 1. Operating Utility-Scale CSP Projects in the United States

Plant	Capacity (MW)	Developer	Technology	Location	PPA With
Kimberlina	5	Areva	Linear Fresnel	Bakersfield, California	Pacific Gas & Electric (PG&E)
Martin Next Generation Solar	75	Florida Power & Light	CSP Trough	Martin County, Florida	Florida Power and Light (FPL)
Nevada Solar One	64	Acciona	CSP Trough	Boulder City, Nevada	PG&E
Sierra SunTower	5	eSolar	CSP Tower	Lancaster, California	Southern California Edison (SCE)
SEGS 1-9	354	Luz International	CSP Trough	Mojave Desert, California	SCE
Total	**503**				

Source: PIX 14955.

Figure 3. SEGS 4, Kramer Junction, California.

CSP plants can be functionally integrated with fossil fuel plants to create hybrid CSP-fossil power plants that can offer peak and base-load power capability. Fossil hybrid plants, also known as integrated solar combined cycle, are under construction in the United States (Florida) and North Africa, including Egypt, Algeria, and Morocco (Richter et al. 2009).

Solar thermal power requires approximately 3–8 acres/MW, depending on the technology and amount of TES. For example, SEGS 3–9 (with a combined capacity of 310 MW) cover more than 1,500 acres, averaging 4.84 acres/MW of gross maximum output (NextEra 2010). In contrast, the Solana station with 6 hours of dispatchable storage will cover approximately 3 square miles, or 6.86 acres/MW of gross maximum output (Solana Solar 2009).

Like other steam-based technologies, CSP (other than parabolic dish) utilizes steam to spin a turbine. Water consumption is a primary consideration for these facilities and can vary from 700–900 gal/MWh, although alternative cooling methods, such as air cooling, can drastically reduce this value at the expense of some efficiency loss and increased cost (Stoddard 2008).

CSP systems are generally classified by the process in which each device collects solar energy. Sections 2.1–2.4 illustrate and compare four primary technologies—CSP trough, CSP tower, parabolic dish, and linear Fresnel reflector. Although only the first two are currently in utility- scale development in the United States, information on CSP-related thermal storage and cooling technologies is also provided.

2.1. CSP Trough

2.1.1. Technology Overview

CSP trough (also referred to as parabolic trough) systems use curved mirrors and single-axis tracking to follow the sun throughout the day, concentrating sunlight on thermally efficient receiver tubes or heat collection elements. A heat transfer fluid (HTF)—typically synthetic oil, molten salt, or steam—circulates in the tubes absorbing the sun's heat before passing through multiple heat exchangers to produce steam. The steam spins a conventional steam cycle turbine to generate electricity or it is integrated into a combined steam and gas turbine cycle when used in hybrid configurations. Utility-scale collector fields are made up of many parallel rows of troughs connected by receiver tubes in series. Rows are typically aligned on a north-south formation axis to track the sun from east to west. Site requirements for a solar trough system include relatively level land, although the solar fields can be divided into two or more terraces. Figure 4 provides a schematic of a CSP trough plant.

Solar troughs are considered the most mature and commercially proven of the CSP technologies. In utility settings, solar trough power plants have shown consistent performance when connected to the electric grid.[9] Improved operating flexibility and dispatchability has been achieved through integration with hybrid fossil systems as well as through demonstrated TES capabilities.

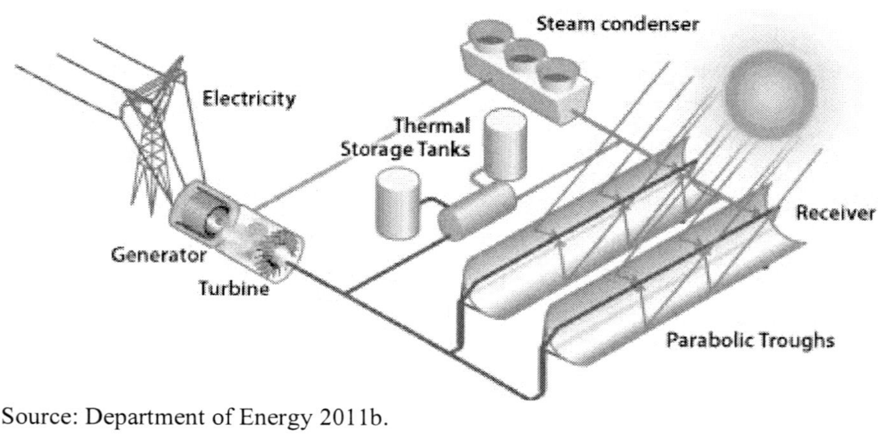

Source: Department of Energy 2011b.

Figure 4. CSP trough schematic.

Source: PIX 16603.

Figure 5. The Nevada Solar One CSP trough system came online in 2007.

There are advantages and disadvantages of different HTFs. Synthetic fuels remain viscous at lower temperatures during the night and on cloudy days but lose efficiency in the heat transfer process (Herrmann et al. 2002). Molten salt, on the other hand, is a highly efficient heat transfer medium that solidifies at lower temperatures. Neither synthetic fuels nor molten salts can directly drive a turbine and therefore must use heat exchangers to boil water and spin a steam turbine. Using steam directly as an HTF is advantageous because it does not require heat exchange equipment; however, it is not very efficient relative to other transfer fluids because it cannot reach high enough temperatures. Further discussion of TES is provided in Section 2.6.

2.1.2. CSP Trough Market Overview

At present, roughly 1,375 MW of utility-scale CSP trough plants are in development with PPAs[10] in place (Table 2). This figure excludes the 100 MW of solar/fossil hybrid plants currently in development. Pacific Gas and Electric (PG&E) holds the majority of trough PPAs, totaling 530 MW.

Solar Millennium made headlines in 2011 when it decided to change its Blythe (1 GW), Amargosa (500 MW), and Palen (500 MW) projects from CSP troughs to PV (PV Magazine 2011; Wesoff and Prior 2011). In doing so, Solar Millennium forfeited a DOE loan guarantee that was acquired to assist development of the Blythe project. Solar Millennium's technology switch was reportedly due to shifting economics as PV modules and other costs have come down in price significantly over the past several years (Clean Energy Authority 2011a). Solar Millennium is currently in insolvency proceedings and has sold its U.S. project pipeline to German developer solarhybrid.[11]

2.2. CSP Tower

2.2.1. Technology Overview

CSP tower systems, often referred to as power towers or central receivers, use a field of mirrors called heliostats that individually track the sun on two axes and redirect sunlight to a receiver at the top of a tower. Sunlight is concentrated 600–1,000 times, making it possible to achieve working fluid temperatures of 500°–800°C (930°–1,470°F) (Australian National University 2010).

Table 2. U.S. Utility-Scale CSP Trough Plants in Development

Plant	MW	Developer	Location	PPA With
Bell Solar Thermal	5	Bell Energy	Tucson, Arizona	Tucson Electric Power
Bethel Energy	50	Bethel Energy, LLC	Imperial Valley, California	San Diego Gas & Electric
Ft. Irwin Solar Power Project	500	Acciona Solar Power	Ft. Irwin, California	U.S. Army
Genesis Solar Energy Project	250	NextEra	Riverside County, California	PG&E
Mojave Solar Power Project	280	Abengoa	San Bernardino County, California	PG&E
Solana Generating Plant	280	Abengoa	Gila Bend, Arizona	Arizona Public Service
Westside Solar Project	10	Pacific Light & Power	Kaua'I, Hawaii	Kaua'I Island Utility Coop.
Total	**1,375**			

Sources: Solar Thermal Magazine 2010; NASDAQ QMS 2006; NREL 2009; CEC 2010b; Solana Solar 2009; Bloomberg 2009; CEC 2010b.

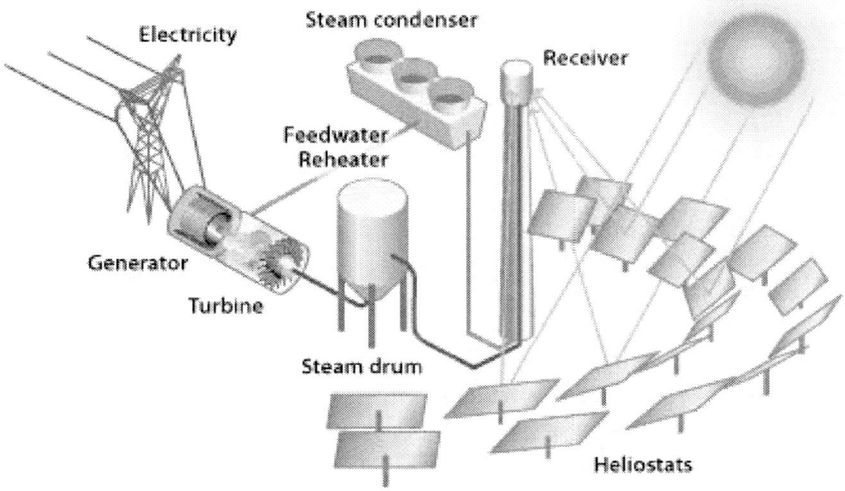

Source: U.S. Department of Energy 2011c.

Figure 6. CSP tower schematic.

Pilot CSP tower plants have proven the technical feasibility of using various HTFs including steam, air, and molten nitrate salts. Early CSP tower systems generated steam directly in the receiver; however, current designs use both steam and molten salt as HTFs. When integrating storage, CSP tower systems have an advantage over CSP troughs since they are able to obtain higher operating temperatures, resulting in a lower required salt inventory for the storage system (Richter et al. 2009).

The largest CSP tower system currently in operation is the PS20 station, designed by Abengoa Solar in Seville, Spain (LaMonica 2009). The 20 MW facility, which began operation in April 2009, features a 531 foot (ft) solar tower and 1,255 heliostats. The PS20 is adjacent to the world's first commercial CSP tower, the PS10, also designed by Abengoa Solar.

In Israel, BrightSource is operating the 4–6 MW Solar Energy Development Center (BrightSource Energy 2011b). According to BrightSource, the facility generates the highest quality steam of any operational solar thermal plant at a temperature of 550°C (1,022°F) and 140 bar (b) pressure.

Also worth noting, the 23 MW Coalinga solar project in central California, recently commissioned in the San Joaquin Valley, utilizes a 327 ft tower system to produce steam (but no electricity) and improve output from an aging nearby oil field. Chevron owns the Coalinga field and the development company that installed the system, Chevron Technology Ventures (IBM 2011).

2.2.2. CSP Tower Market Overview

Some 2,655 MW of proposed CSP tower systems are currently under contract with U.S. utilities. BrightSource Energy has the most megawatts under contract. In April 2011, BrightSource closed on a $1.6 billion DOE loan guarantee for its Ivanpah, California, facility (DOE 2011c). Of BrightSource's 2.2 GW portfolio under contract, Ivanpah represents 392 MW, which allocated about evenly between PG&E and SCE. Many of BrightSource's other projects are at undisclosed locations. In October 2010, BrightSource broke ground on the Ivanpah project and received a $300 million investment from NRG Energy. With this investment, NRG Energy will hold a majority equity stake in the project (Murray 2010).

One small utility-scale CSP tower system operates in the United States—eSolar's 5 MW Sierra Suntower. The facility became operational in 2009 and sells power to SCE (NREL 2010b). In co-development with NRG Energy, eSolar has two proposed facilities, the Gaskell Sun Tower phases 1 and 2, under long-term contracts with IOUs for a total of 245 MW. To help lower costs, eSolar deploys a modular design surrounding a conventional turbine (eSolar 2010).

Source: PIX 00036.

Figure 7. The Solar One facility in California employed CSP tower technology.

Table 3. U.S. Utility-Scale Central Receiver Projects in Development

Plant	MW	Developer	Location	PPA With
BrightSource, PG&E PPA	108	BrightSource	California	PG&E
Coyote Springs 1 & 2	400	BrightSource	Coyote Springs, Nevada	PG&E
Crescent Dunes	110	SolarReserve	Nye County, Nevada	NV Energy
Gaskell Sun Tower (Phases 1 & 2)	245	NRG/eSolar	Kern County, California	SCE
Hidden Hills 1 & 2	500	BrightSource	Inyo County, California	PG&E
Ivanpah Phases 1–3	392	BrightSource	Ivanpah, California	PG&E
Rice Solar Energy Project	150	SolarReserve	Blythe, California	PG&E
Rio Mesa 1–3	750	BrightSource	Riverside County, California	SCE
Total	**2,655**			

SolarReserve has two CSP tower facilities under development—Crescent Dunes and Rice Solar Energy Project—totaling 260 MW and 25-year contracts with PG&E and NV Energy (Reuters 2009). SolarReserve was founded by United Technologies Corp., whose Rocketdyne subsidiary demonstrated the solar tower technology at the Solar One and Solar Two power plants in southern California. However, both facilities were demonstration projects and are no longer operating (Solar Reserve 2010). U.S. Renewables Group, a large private equity firm exclusively focused on clean fuel projects, supports SolarReserve (SolarReserve 2011).

2.3. Parabolic Dish

2.3.1. Technology Overview

Parabolic dish, or dish engine, systems are individual units comprised of a solar concentrator, a receiver, and an engine or generator. The concentrator typically consists of multiple mirror facets that form a parabolic dish, which tracks the sun on two axes and redirects solar radiation to a receiver (Richter et al. 2009). The receiver is mounted on an arm at the focal point of the reflectors and contains a motor-generator combination that operates using either a Stirling engine or a small gas turbine. Dish systems are generally between 10 kilowatts (kW) and 25 kW in size. Compared with other CSP technologies, parabolic dish conversion efficiencies are the highest, reaching over 30% (SolarPACES 2010).

Parabolic dish systems are considered highly modular, allowing individual deployment for remote applications or groupings for small-grid or large-scale utility applications (SolarPACES 2010). Individual placement also enables greater flexibility than other CSP systems since dish systems can be placed on varied terrain with grades up to 5% (TEEIC 2010). In addition, parabolic dish technology only uses small quantities of water, mostly for washing concentrators free of dust. However, due to current economies of scale, dish systems are generally only proposed in utility-scale projects.

2.3.2. Parabolic Dish Market Overview

At present, there are no utility-scale parabolic dish projects in development.[12] Through 2010, one company—Tessera Solar—held at least three contracts with western U.S. utilities, representing more than 1,600 MW. Tessera was the development affiliate to Stirling Energy Systems, which was a manufacturer of parabolic dishes and Stirling solar engines before filing for Chapter 7 bankruptcy in 2011 (Wesoff 2011).

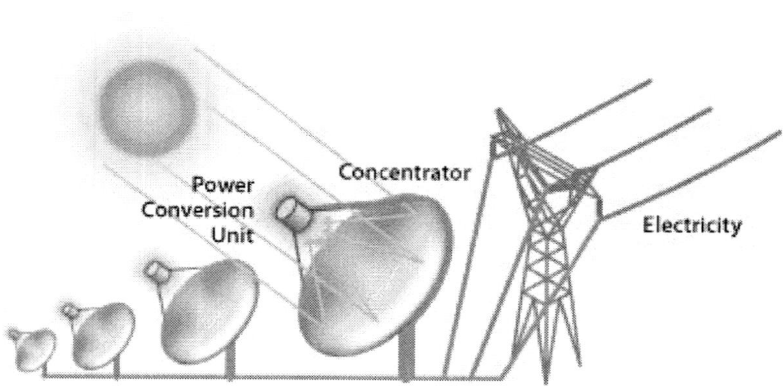

Source: DOE 2011d.

Figure 8. Schematic of a parabolic dish system.

In May 2011, Tessera lost its last contract when the developer that bought the project, AES, decided to replace the parabolic dish technology with PV. Greentech Media reported that Tessera could not secure a DOE loan guarantee and was thus unable to fulfill the contract (Wesoff 2011).

2.4. Linear Fresnel Reflector

2.4.1. Technology Overview

Linear Fresnel reflector, also referred to as compact or concentrating linear Fresnel reflector, systems are made up of flat or nearly flat mirror arrays that reflect solar radiation onto elevated linear absorbers or receiver tubes. Water, the typical thermal fluid, flows through the tubes and is converted into steam. Steam can also be generated directly in the solar field, eliminating the need for costly heat exchangers (DOE 2010b). The system is similar to a CSP trough in that the sunlight is concentrated in a linear fashion. However, instead of a single curved mirror, linear Fresnel systems concentrate the insolation of many slightly curved mirrors onto a receiver. The receiver is stationary and does not move with the mirrors as in the CSP trough systems, so it does not require rotating couplings between the receivers and the field header piping, thus providing additional design flexibility.

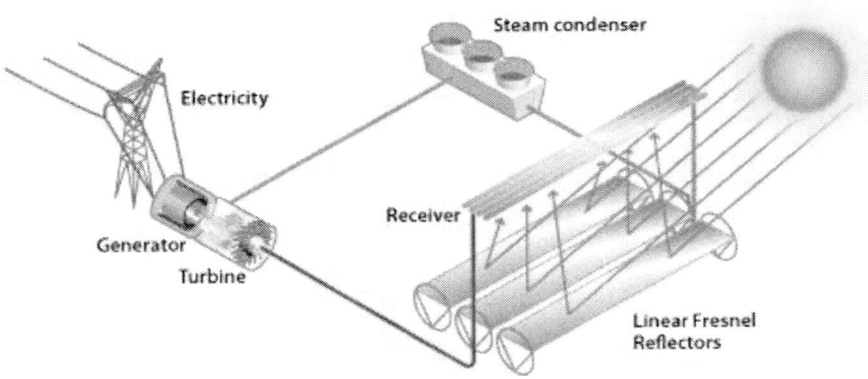

Source: DOE 2011c.

Figure 9. Linear Fresnel reflector schematic.

2.4.2. Linear Fresnel Reflector Market Overview

In 2010, Ausra—the sole developer of linear Fresnel projects in the United States—sold its technology and development pipeline to the French company Areva (Baker 2010). To date, Areva's 5 MW Kimberlina project in Bakersfield, California (previously developed and owned by Ausra), is the only utility-scale linear Fresnel reflector project in the United States. Prior to the Areva sale, Ausra was developing the Carrizo Energy Solar Farm, a 177 MW project, but that project was suspended.

2.5. Solar-Fossil Hybrid Power

2.5.1. Technology Overview

Hybrid power plants incorporate both solar collector fields and fossil fuel combustion to generate power, often relying on a common steam cycle and allowing for power production during sunlight fluctuations and nighttime hours.[13] There are many variations of hybrid plants, including simple natural gas backup, integrated solar combined cycle plants, and solar plants providing thermal input to existing or newly designed coal-fired plants. To produce steam in hybrid plants, CSP trough, CSP tower, and linear Fresnel collector devices may be used. Figure 10 is a rendering of a solar-fossil (gas turbine/CSP trough) hybrid facility.

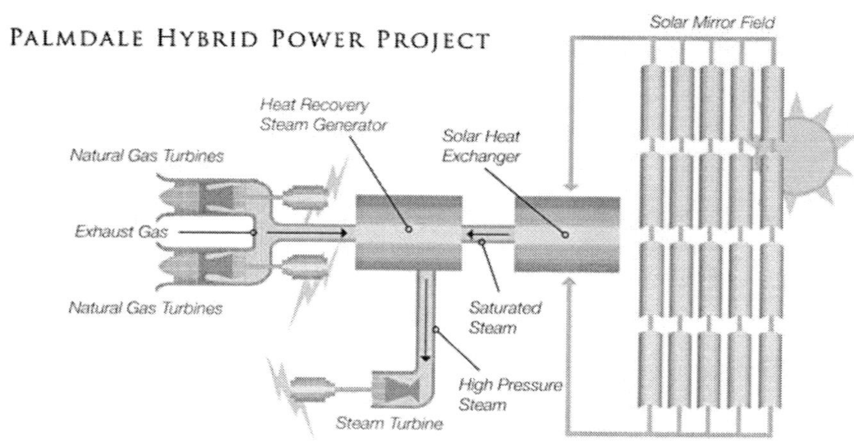

Source: Inland Energy 2011.

Figure 10. Rendering of a solar/fossil hybrid facility.

Combining CSP and fossil fuel power is not a new concept. In fact, many CSP plants use natural gas as a backup energy source. Assuming space requirements are adequate, it is possible to retrofit existing power plants with solar thermal technology, an option that may be advantageous for utilities looking to increase the efficiency of their fleets. By combining the components of technologically proven fossil fuel plants with the environmental benefits of CSP, there could be an increase in market opportunities and competition with conventional power plants.

2.5.2. Solar/Fossil Hybrid Market Overview

One solar/fossil electric generating plant, as defined by NREL, is currently in operation—the Martin Next Generation Solar Energy Center. The plant combines 75 MW of CSP trough with a 3,705 MW natural gas- and oil-fired generation facility.

As shown in Table 4, two utility-scale solar/fossil hybrid plants are currently in development, the Palmdale and Victorville 2 projects. These two plants feature similar hybrid designs including CSP trough and combined cycle technology designed and constructed as a combined facility (Inland Energy 2011). In each project, the solar field will provide approximately 10% of the thermal input. Both projects are also proposed to be constructed and owned by municipalities. The Victorville 2 project was approved by the California Energy Commission (CEC) in 2008 (City of Victorville 2008). In August 2011, the CEC formally approved development of the Palmdale project (CEC 2011).

A large solar hybrid project, the San Joaquin 1 and 2 facilities, was recently cancelled due to "issues regarding project economics" and other aspects of the project (Martifer Renewables 2010). Additionally, the 4 MW Cameo hybrid demonstration project in Grand Junction, Colorado, was recently decommissioned and dismantled. Cameo was the first power plant to hybridize solar troughs and coal-fired generation.

2.6. Thermal Energy Storage

TES provides the ability of a system to store thermal energy collected by a solar field in a reservoir for conversion to electricity at another time. For CSP technologies, storage can be used to balance energy demand between day and night or during times of intermittent sunlight. By oversizing the solar fields and pulling the excess heat to the thermal storage component, the turbine can operate at a fairly constant rate. Figure 11 illustrates this process.

A storage system enables CSP plants to (1) negate the variability in system output due to sudden shifts in the weather and (2) extend the range of operation of a CSP system beyond daylight hours (Biello 2009). The power produced throughout the day can be more effectively matched with energy demand, therefore increasing the value of the power as well as the total useful power output of the plant at a given maximum turbine capacity.

A well-located CSP trough plant with no fossil backup or thermal storage should be able to achieve a 25% annual capacity factor (NREL 2011a). CSP with storage is theoretically capable of capacity factors around 75%, although economic application of storage limits the capacity factor to approximately 50% given current available technology.[14] CSP generation facilities supported through the DOE loan guarantee program have capacity factors that range from 26%– 28% for projects without thermal storage to 43%–52% for projects with thermal storage (DOE 2011c).

2.6.1. Technology Overview

Storage mechanisms are classified as either direct or indirect based on how the storage medium is heated by the solar concentrators. Indirect systems, such as most CSP trough plants, use a separate HTF, such as synthetic oil, that passes through a heat exchanger to heat the storage medium. Direct systems use the same fluid, such as steam, for both the HTF and the storage fluid eliminating the need for expensive heat exchangers.

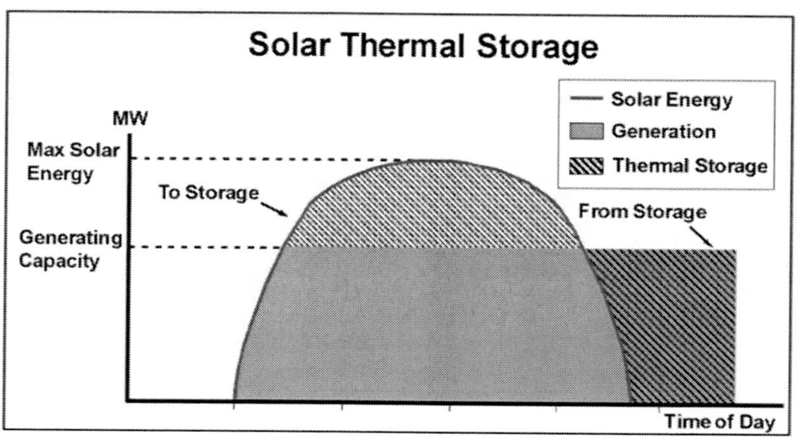

Figure 11. Solar thermal storage extends the power production period.

Table 4. U.S. Utility-Scale Solar-Fossil Hybrid Projects under Development

Plant	Solar/ Total MW	Developer	CSP and Fossil Technology	PPA With
Palmdale Hybrid Power Project	50/570	Contractor not selected yet	CSP trough/ natural gas combined cycle	City of Palmdale
Victorville 2 Hybrid Power Project	50/513	Contractor not selected yet	CSP trough/natural gas combined cycle	City of Victorville
Solar Total	**100/1,083**			

Source: PIX 02185.

Figure 12. The Solar Two system in California included a thermal energy storage system.

Molten salt storage systems, which can be used in direct or indirect storage systems, seem to hold the greatest promise of economic commercialization (Price 2009). Molten salt systems allow the solar field to operate at higher temperatures relative to other fluids or storage media, reducing the cost of the system. Because salts melt at very high temperatures (e.g., ordinary table salt melts at around 1,472°F), they can hold significant quantities of heat without vaporizing (Biello 2009). A mixture of sodium nitrate and potassium nitrate, the salts can efficiently return as much as 93% of the energy sent into storage.

However, a technical disadvantage of molten salts is that they freeze at relatively high temperatures, from 120°–220°C (250°–430°F). Sandia National Laboratories is currently developing new salt mixtures with the potential for lower freezing points below 100°C (212°F) to help solve this challenge (Biello 2010).

2.6.2. Thermal Energy Storage Market Overview

TES offers potential long-term cost advantages for CSP plants by amortizing the fixed cost of the power block over greater electricity generation. However, a lack of development and operational experience has limited technology use to date.

The Andasol plant in Spain, developed by Solar Millennium, utilizes 28,500 metric tons of molten salt to provide 7.5 hours of backup generation at full output (Solar Millennium 2010). The salt utilized in the plant is 60% sodium nitrate and 40% potassium nitrate, both commonly found in fertilizers and other materials.

In the United States, no operating CSP plants utilize thermal storage, although several are in development. Abengoa Solar's Solana power station is expected to store 6 hours of thermal energy (NREL 2010a). Located outside Gila Bend, Arizona, the 250 MW (net)[15] facility is projected to cost $2.00 billion, $1.45 billion of which will be paid for with debt financing covered under a DOE loan guarantee (Prior 2010). In late 2011, BrightSource announced that it will add storage capability to three of its PPAs with SCE (BrightSource Energy 2011a).[16]

Bell Independent Power Corporation (Bell) is also developing a CSP and combined thermal storage facility. The 5 MW plant will be part of the new Tech Park in Tucson, Arizona (Environmental Leader 2010), and was the result of Bell's request for proposal submission to Tucson Electric Power (TEP). Bell and TEP signed a 20-year contract, which is currently awaiting approval from the Arizona Corporation Commission (Solar Thermal Magazine 2010). The facility is expected to begin operating in 2012.

A Discussion on Capacity Factors

Capacity factor is the ratio of actual output of power over a period of time compared to the output of full nameplate capacity operation. Solar technologies have relatively low capacity factors because they only produce power when the sun is shining. Other technologies, such as coal or natural gas, can produce power at a relatively constant rate or as dictated by demand.

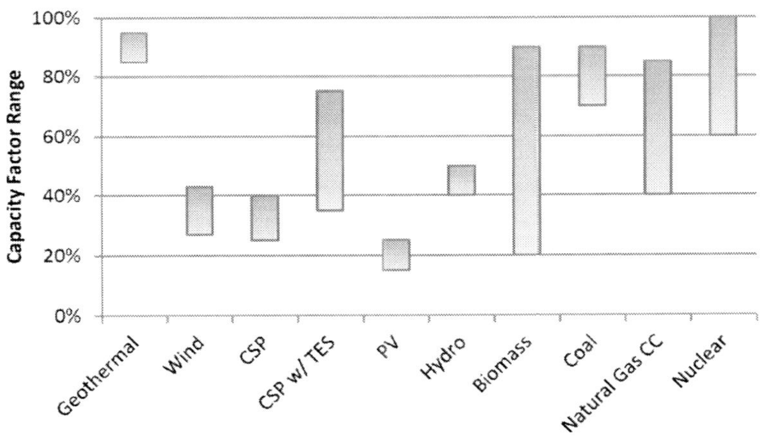

Figure 13. Comparison of capacity factor by technologies.

The capacity factor for PV technologies ranges from 14%–18% for thin-film systems and 20%–24% for crystalline installations. Thermal storage can significantly increase the capacity factor of eligible CSP plants from 25% without storage to approximately 75% with storage.

Source: Renewable Energy Research Laboratory 2011.

2.7. Cooling Systems

Steam-driven power plants, such as CSP facilities, require a consistent source of fresh water, which can be difficult to obtain in the desert where the solar resource is plentiful. Water consumption is primarily connected to the cooling system. There are three primary types of cooling systems: open loop, closed loop, and dry. Open loop, or once-through, cooling systems pull heat from the power plant by withdrawing large quantities of water from rivers and other sources and returning the now-warmer water to its source. As most of the water used in an open loop system is returned to its source, these systems actually consume (via evaporation) very small quantities of water (DOE 2008).

Due to environmental concerns associated with increasing the temperature of river water, open loop systems were disallowed in new power generation facilities in the early 1970s (California Environmental Protection Agency 2008). Nonetheless, open loop cooling systems grandfathered into the new law are still widely used throughout the United States. According to the DOE, more than half of the existing fleet of thermal generating plants in the United States are estimated to be equipped with once-through cooling systems (DOE 2008). The U.S. Environmental Protection Agency is in the process of developing new rules associated with Section 316(b) of the Clean Water Act that will help determine when open loop cooling will be allowed. Recent revisions to the draft rules gave power developers more flexibility in water cooling, although some may still need to switch from open to closed loop systems.

Closed loop cooling systems cool and recirculate water within the power plant and thus withdraw far less water than open loop systems. However, during the cooling process, water is lost via evaporation. Closed loop systems negate thermal pollution of water sources and withdraw far less water but lower plant efficiency by approximately 0.8%–1.4% (DOE 2008).

Dry cooling, or air cooled, systems use air to condense heat and cool power plants. These systems have minimal water requirements—either in withdrawal or consumptive modes—and can generally be used in all steam cycle power plant technologies, including CSP trough and CSP tower facilities (DOE 2008). However, dry cooling systems are more expensive to build and can lower the efficiency and output of the power plant, especially on very hot days.

To help balance cost, plant output, efficiency, and water use, some power plants are being designed with hybrid cooling systems that combine closed loop wet and dry cooling systems (DOE 2006). Air cooling dissipates heat directly into the air, using water only for general plant uses and steam cycle blowdown, which eliminates dissolved solids in the steam. Hybrid cooling systems can reduce water use by 50%–85% with only a 1%–3% drop in power output (DOE 2010a).

Table 5. Water Usage Requirements for Electric Generation Technologies

Cooling System	Power Plant Technology	Water Usage (Gallons/MWh)	
		Withdrawal	Consumption
Open loop	Fossil/biomass waste/nuclear steam	20,000–60,000	100–400
	Natural gas combined cycle	7,500–20,000	50–100
Closed loop	Fossil/biomass waste steam	300–600	300–1,100
	Nuclear	500–1,100	700–850
	Geothermal	2,000	1,400
	Solar trough	760–920	720–1,050
	Solar tower	750	740–850
Dry	Various technologies	0	0–80
Hybrid	Various technologies	50–650	100–600

Source: DOE 2008; Macknick et al. 2011.

Source: PIX 15558.

Figure 14. SunEdison's 8.2 MW Alamosa plant.

3. PHOTOVOLTAIC SOLAR POWER

In the last several years, solar PV has achieved unprecedented scales of deployment. A range of factors have contributed to this development, including a rapid decline in solar module prices, increased requirements on utility procurement of renewable energy via state RPS programs, and financial incentives available through the federal government. Total installed PV capacity has increased from 22 MW in 2000 to nearly 3.1 GW as of the third quarter of 2011 (SEIA 2011b). Over 1 GW of grid-tied PV was added in the first three quarters of 2011 alone.

Utility-scale installations have seen the greatest growth in the last two years, increasing from 70 MW installed in 2009 to over 700 MW installed in 2011 (SEIA 2011b; SEIA 2012). In 2011 alone, utility installations were up 185% from the previous year. Going forward, NREL estimates that there are more than 11 GW of utility-scale PV projects in the pipeline with signed PPAs as of January 2012.

3.1. Traditional PV

3.1.1. Technology Overview

PV systems consist of c-Si or thin-film solar modules. c-Si materials include monocrystalline or polycrystalline cells. Thin-film PV includes an array of materials including CdTe, copper indium gallium selenide (CIGS), and a-Si. These materials are generally less expensive to produce than c-Si but have lower conversion efficiencies.[17] Although many thin-film materials have been introduced over the past several years, steep price reductions in c-Si modules have overridden the technology's cost advantage.

The majority of PV modules—about 80%–90%—currently installed in the United States are based on c-Si wafers (Solarbuzz 2010).[18] Silicon wafers are produced either by slicing sections of bulk silicon or growing thin ribbons that are sectioned into wafers (Bates 2000). Since silicon has a high melting point, the production process is energy intensive. In addition, much of the bulk material is lost when cutting the wafers.

Several companies have attempted to innovate on the process of silicon production to eliminate this waste and hasten the production process as well. For example, 1366 Technologies, a solar cell manufacturer based in North Lexington, Massachusetts, has devised a continuous process that uses a molten silicon bath to form wafers instead of the traditional ingots and boules. The company secured a $150 million loan guarantee from the DOE in September 2011 to scale up manufacturing and had secured two prior DOE grants to pioneer the technology. The molten silicon process is estimated to achieve production cost reductions of 50%, energy use reductions of 90%, and the generation of silicon wafers in a fraction of the time of current production standards (DOE 2011a).

Before the end of 2008, the demand for crystalline solar panels outpaced the industry's ability to produce them, a trend that opened the door for other solar module alternatives, such as thin films (Wang 2009b). Manufacturing c-Si has since ramped up, and as of 2011 it appears to have outpaced demand.

DERATE FACTORS FOR PHOTOVOLTAIC SYSTEM COMPONENTS

The overall DC-to-AC derate factor varies for different PV systems and applications. NREL's PVWatts tool incorporates a standard derate factor of 0.77 (or a 23% loss in output from nameplate DC rating to actual AC energy produced).

Table 6. PVWatts Default Derate Values

Component Derate Factors	PVWatts Default	Range
PV module nameplate DC rating	95%	0.80–1.05
Inverter and transformer	92%	0.88–0.98
Mismatch	98%	0.97–0.995
Diodes and connections	100%	0.99–0.997
DC wiring	98%	0.97–0.99
AC wiring	99%	0.98–0.993
Soiling	95%	0.30–0.995
System availability	98%	0.00–0.995
Shading	100%	0.00–1.00
Sun-tracking	100%	0.95–1.00
Age	100%	0.70–1.00
Overall DC-to-AC deratefactor	77%	0.09999–0.96001

Source: NREL 2011b.

PV arrays require inverters to convert direct current (DC) power produced by the modules into alternating current (AC), which can then be connected to the electrical grid. Throughout the components of the system there are electrical losses, which derate the conversion from nameplate DC power rating to AC power rating (as explained in Table 6). Row spacing is very important in order to maximize power density (kW/acre) and minimize panel shading. If only a portion of a cell, panel, or array is shaded, the power output can fall—sometimes dramatically.[19] In contrast to CSP technologies, which require direct sunlight to operate properly, PV systems utilize both direct and indirect (diffuse) radiation (Pitz-Paal 2009). For this reason, PV systems can be more widely deployed and can even operate on cloudy days (albeit at a lower capacity). Under cloudy weather and dawn and dusk conditions, thin-film solar panels convert low and diffuse sunlight to electricity more efficiently than crystalline-based panels. Current PV conversion efficiencies vary by technology and manufacturer, though generally speaking, thin-film materials have lower efficiencies than silicon wafers. First Solar, who deals exclusively in CdTe modules, reported an average efficiency of 11.7% in 2011 (up from 11.4% in 2010) and recently announced a record-setting 14.4% for their highest efficiency module (First Solar 2012). Modules in the c-Si space run between 14% and almost 20%[20] efficiency in the field, with 23% efficiency achieved in controlled laboratory testing (referred to "champion- production" efficiency) (Green et al. 2011).

PV systems do not require water to operate. However, a small amount of water is needed to clean the panels and is recommended in areas that have little rain. Operation and maintenance costs are minimal for these systems because there are few moving parts and no need to service turbines or generators (see Figure 17). The largest cost of PV technology is for the modules, approximately 50% of the total, followed by the installation materials, labor, and the inverters. Inverter replacement can be a significant expense. PV module warranties are generally 20–25 years long; however, inverter warranties are typically 10–15 years long (Russell 2010). Technological improvements are occurring rapidly in many subsectors. For example, microinverters can be paired with each PV module, in contrast to centralized inverters, which are paired with a bank of modules. Therefore, if a single microinverter fails, only the module paired to the failed inverter is affected (Russell 2010).

There is concern about the life of both microinverters and centralized inverters. Improvements are being made in this area, too. For instance, in October 2010, SolarBridge Technologies announced a 25-year warranty on its new microinverter, the SolarBridge AC Module System (SolarBridge 2010).

3.1.2. PV Market Overview

The PV industry's extraordinary growth has not been without challenges. From the overheating and contraction of the Spanish market in 2007 and 2008 (the result of miscalculations in their feed-in tariff program) to plummeting module prices, the solar industry has proven a very dynamic market. Many companies have downsized or withdrawn completely from the market. Between late 2008 and 2011, several PV manufacturers and developers, including OptiSolar, SunPower, SunTech Power Holdings, Energy Conversion

Devices, Solyndra, and SunEdison, laid off workers or shuttered operations (Baker 2009; Fehrenbacher 2009; Wang 2010).

Despite this challenging landscape, solar industry job growth in the United States has been significant over the last several years. The most recent *Solar Jobs Census* estimates that as of August 2011, there were 100,237 solar employees (workers who spend over 50% of their time supporting solar-related activity), which is up 6.8% from August 2010 (Solar Foundation 2010).

To analyze the utility-scale market, this report segments PV into three general project sizes. Systems smaller than 20 MW are commonly referred to as "distributed" systems, as they often tie into distribution level voltages.[21] Systems 20–49 MW are referred to as "mid-sized," and those 50 MW and larger are referred to as "large-sized."

3.1.2.1. Distributed Utility-Scale PV Systems

Distributed systems are often integrated into the distribution grid interconnecting below 69 kilovolts, which is generally considered the level at which transmission service begins (Wisconsin Public Service Commission 2009). However, for several reasons these systems are applicable for consideration as utility-scale systems and were therefore included in this analysis. At this project size, power is almost exclusively sold to utilities. Most customers do not have the load to support 5 MW or more of on-site PV.[22] In addition, in order for solar projects to access relatively low-cost capital, the PPA contract must be signed with a counterparty that has an excellent credit rating (Feo and Tracy 2009). Utilities are generally considered excellent counterparties due to their bonding capability and regulatory mission to provide power to their customers.[23] Appendix Table B-1 represents all distributed utility-scale (5.0–19.9 MW) projects currently under development that hold PPAs. PV project development at the distributed utility-scale has many advantages. Environmental permitting, grid integration, and site control can be less burdensome at this project size. Certain companies are also considering development at brownfield sites, including former mining operations and other disturbed lands, which can further reduce environmental oversight of project development (Mendelsohn 2010).

In California, generators up to 20 MW are subject to the Small Generator Interconnection Procedures (SGIP), which avoids the need to be grouped into a "cluster" for a time-consuming, grid-wide system impact study. However, this shortcut has attracted a lot of projects just under the threshold, requiring the California independent system operator (ISO) to reassess the SGIP (Mendelsohn 2010).

According to NREL's internal database, approximately 65 distributed projects totaling 589.2 MW of capacity are currently under development and hold PPAs. See Appendix Table B-1 for information on these projects.

COST OF PV FACILITY BY COMPONENTS

Figure 15 shows the cost breakdown for a fixed-tilt utility-scale PV system utilizing c-Si modules. Lower efficiency thin-film modules generally cost less but can have higher balance of plant (or non-module) expenses. This includes costs for supporting structures, DC cabling, and inverters.

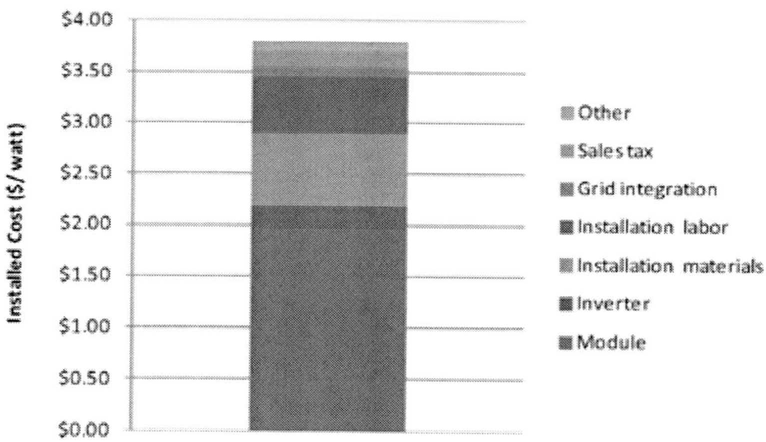

Source: Goodrich et al. 2012.

Figure 15. Utility-scale PV facility by cost component.

3.1.2.2. Mid-Sized Utility-Scale PV Systems

In the past few years, many announcements regarding PV systems in the 20–49 MW range have been made. The first two systems in this size range were completed in 2009: the 25 MW DeSoto and the 21 MW Blythe projects. Prior to these projects, the largest system in the United States was the 14 MW Nellis Air Force Base system. As shown in Appendix Table B-2, there are currently 57 mid-sized projects in advanced development (i.e., with PPAs in place) totaling 1,329.5 MW of capacity. SCE leads all utilities in this size category with 18 contracts totaling 384 MW. See Table B-2 for more information on these projects.

According to NREL's database of project announcements, First Solar deploys its own CdTe thin- film panels in its projects. First Solar CdTe modules have been selected for other projects, including juwi solar's BlueWing facility in Texas (Vo 2010). Most other projects are expected to utilize c-Si panels, although the panel manufacturer and even the PV material are not always disclosed as part of the project announcement. In part, this may be because developers wait to procure the lowest-cost components available until just prior to installation.

3.1.2.3. Large-Sized Utility-Scale Systems

According to NREL's utility-scale projects, there are currently 9,425 MW of large-sized (greater than 50 MW) PV projects currently holding a PPA with a utility purchaser (see Table B-3). None are currently operating. First Solar is the clear leader among developers with 1,920 MW of capacity in the development or construction phase. This amounts to about one-fifth of the total capacity of large-sized projects under development. First Solar also has one of the largest projects under current development, the 550 MW Topaz Solar Farm, which will sell its energy to PG&E (Wang 2009). For purposes of comparison, the Topaz project is 10 times larger than the largest PV project completed in the United States: the recently completed 55 MW Copper Mountain Solar Facility in Nevada (which also uses First Solar's CdTe panels). SunPower is another player with a significant stake in the large-sized solar project space. The developer signed three PPAs with SCE in 2011 for the sale of 711 MW of power from three facilities. These three projects represent all of SunPower's plays in the large-sized category.

Energy developer enXco recently signed an agreement with Japanese manufacturer Solar Frontier to deliver up to 150 MW of CIS[24] panels for use in enXco's PPAs with SDG&E. These will be the first CIS modules deployed at utility scale in the United States.

Development on Public Lands

Many large-scale solar projects, including those without PPA contracts, have been proposed for development on federal land administered by the U.S. Department of the Interior's Bureau of Land Management (BLM). Interior Secretary Ken Salazar announced several initiatives to improve the speed of BLM application review.

In 2009, there were 470 applications, with 158 of those being solar, to develop 97,000 MW of renewable energy projects on BLM land (U.S. Department of the Interior 2009). Federal agencies were to work with western leaders to designate tracts of BLM land as suitable for large-scale solar energy development, fund environmental studies, open new solar energy permitting offices, and speed reviews of industry proposals (U.S. Department of the Interior 2009).

For several years, the BLM has been involved in a wide-scale environmental evaluation of solar energy development across BLM lands. The process, referred to as the Solar Programmatic Environmental Impact Statement (PEIS), was expanded to review 24 solar energy study areas. In October 2009, the BLM issued over 1,100 pages of comments that will be considered when developing the draft PEIS (BLM 2010).

In June 2010, the BLM issued a guidance document on the rental fees to be applied to solar facilities for use of BLM land. The BLM will apply two separate rental fees: (1) base rent, which is applied on a per-acre basis to specific counties and ranges from $15.70/acre/year to $313.88/acre/year; and (2) megawatt capacity fee, which is applied on a per-megawatt basis and ranges from $5,256/MW/year for PV projects to $7,884/MW/year for CSP with 3 hours or more of thermal storage (BLM 2010). In January 2011, BLM issued a solar energy plan of development, which provides guidance on submitting solar applications to the agency, including outlining expected information with respect to facility engineering, project construction, and transmission interconnection (BLM 2011).

One project of note on BLM land is BrightSource Energy's Ivanpah Solar Project. The Ivanpah site is home to the desert tortoise, classified as "threatened" on the endangered species list (Wesoff 2011b). Construction at the Ivanpah site was halted in April 2011, and after further analysis, was allowed to resume in June 2011. According to a report submitted to the BLM and the CEC, 20 live tortoises were observed to be living within the project boundary (Solar Power Partners 2010). BrightSource was required to mitigate the environmental impact of the project by significantly reducing the overall footprint. Environmental oversight will remain over the project's life and, likely, beyond.

Many more utility-scale solar projects than those listed here have been proposed but do not necessarily have a PPA contract in place. California alone has tabulated over 70,000 MW of proposed renewable energy projects for the state, over 90% of which are projected to be 200 MW and larger (CEC 2010b).

3.2. Concentrating PV

3.2.1. Technology Overview

CPV systems use optic lenses to focus sunlight onto high-efficiency solar cells. Due to the integration of lenses, CPV systems require direct sunlight to operate, so most systems employ single- or dual-axis trackers to follow the sun across the sky (BLM 2010).

CPV technologies combine two components to generate electricity: concentrators, which can be either lenses or mirrors, and a semiconducting material based on the array of available PV technologies. Depending on a selected strategy, a manufacturer will employ a combination of concentrators and semiconducting material to achieve the lowest cost of energy.

The potential cost advantage of CPV over traditional PV is the use of less semiconducting material that comprises the cell. Lower efficiency CPV technologies may employ silicon (CdTe and CIGS can also be used), but the highest efficiencies are achieved with multi-junction cells using materials other than silicon. Field efficiencies for these cells are in the 30% range and laboratory tests have achieved upwards of 40% efficiency (Kurtz 2011).

Current concentration intensities range from as low as 2x (or 2 "suns") to as high as 500x. Some emerging technologies are anticipated to reach up to 1,000x. CPV concentrations are often distinguished as low, medium, and high as roughly outlined in Table 7. Because of the high temperatures inherent to light concentration, CPV systems often require integration of heat sinks to dissipate heat and prevent damage to the solar cells.

CPV systems function similarly to PV systems in that they utilize inverters to convert the DC generated by the solar cells to AC, which can then be delivered to the electrical grid. These systems have low water requirements and, due to their high power density, have lower land requirements than other solar technologies. In addition, CPV technologies have the potential to significantly reduce costs because the majority of the system expense is in the lenses, mirrors, and tracking equipment—components that are highly susceptible to economies of scale.

3.2.2. CPV Market Overview

CPV systems are the least commercially deployed of the PV technologies. The only utility-scale plant currently in operation is a 5 MW project located in Hatch, New Mexico, commissioned in June 2011. NextEra Energy Resources owns the project, which uses Amonix modules—high- efficiency, multi-junction gallium indium PV cells covered with Fresnel lenses (Clean Energy

Authority 2011b). Amonix is one of two manufacturers currently supplying technology to the utility-scale CPV market in the United States; the other is French semiconductor company Soitec.[25] All systems are considered high concentration.[26]

The limited commercial success of CPV to date is partly due to the fact that these systems are more complex than PV systems. During 2008, as silicon prices were reaching new market highs, CPV systems appeared ready for a commercial breakthrough (Barron 2008). Prices have since collapsed, however, and this has changed the economics of several alternative technologies, including CPV (Hamberg 2011).

Table 7. CPV Concentration Classes

Class of CPV	Typical Concentration Ratio	Type of Converter
High-concentration	>400X	Multi-junction
Medium-concentration	~3X–100X	Silicon or other cells
Low-concentration	<3X	Silicon modules

Table 8. CPV Systems under Development

Facility Name	Capacity (MW)	Developed By	Location	Power Sold To
Alamosa Solar Generating Project	30	Amonix/ Cogentrix	Near Alamosa, Colorado	Xcel Energy
Desert Green Solar Farm	6.5	Soitec	Borrego Springs, California	SDG&E
Imperial Solar Energy Center West	250	Tenaska Solar Ventures	Imperial County, California	SDG&E
LanEast Solar Farm	22	Soitec	Boulevard, California	SDG&E
LanWest Solar Farm	6.5	Soitec	Boulevard, California	SDG&E
Littlerock Solar Power Generation Station	5	Amonix	Torrance, California	SCE
Lucerne Solar Power Generation Station	14	Amonix	Lucerne Valley, California	SCE
Rugged	80	Soitec	Boulevard, California	SDG&E
South Swan Solar Project	12	Amonix	Swan Road Quarry, Arizona	TEP
Total	**471 MW**			

Source: PIX 19186.

Figure 16. CPV modules at the SolarTAC testing facility in Aurora, Colorado.

Despite the dramatic decreases in silicon and conventional module pricing, the CPV market looks to be entering a tentative growth stage. According to NREL's database, at least 10 utility- scale CPV projects representing about 471 MW are currently in development and hold long-term PPAs with utilities. San Diego Gas and Electric (SDG&E) holds the majority of these PPAs, both in terms of megawatts (410 MW, or 86% of total) and absolute numbers (6).

One CPV project, the 30 MW Alamosa Solar Generating Project in Colorado, will be the largest CPV installation in the world when completed in 2012 (Pankratz 2011).[27] Project developer Cogentrix received a DOE loan guarantee of $90.6 million in September 2011; this was the only loan guarantee awarded to a CPV project.

Continued market growth for CPV will be the most important factor in keeping its costs competitive with traditional PV and with fossil fuels. Without manufacturing in the tens of megawatts per year, it is unlikely that CPV will achieve the cost reductions necessary to make it an economic technology, despite its high efficiencies (Kurtz 2011).

4. Summary of Market Highlights

Utility-scale solar capacity additions have grown exponentially in the last several years, and the development pipeline ensures that significant growth will maintain through the near-term. Plunging PV costs have altered the competitive landscape amongst technologies but have generally improved project economics enabling large-scale solar deployment. Key observations from this report include:

- PV leads the utility-scale solar market for projects under development with approximately 72% of the capacity under long-term contract. In contrast, only about 57% of the current operational capacity in the United States derives from PV technologies. About 71% of PV projects under development have indicated or are expected to utilize c-Si modules made from a wide variety of manufacturers; approximately 23% are contracted to use CdTe modules, most of which will be manufactured by First Solar.
- California utilities are priming the market by signing large PPA contracts with utility- scale developers. Combined, California's three IOUs represent 72% of the total utility-scale market in the United States.
- Though CSP trough plants dominate the CSP market today, there is nearly half as much CSP trough capacity in development as that for CSP tower. The contraction of the CSP market is largely due to the exit of Solar Millennium, who announced in late 2011 that they would be switching the technology for their Blyth and Palen facilities from troughs to PV.
- Conversely, CSP tower technologies have risen to prominence in the U.S. development pipeline, representing approximately 2,655 MW, or 16% of planned utility-scale solar capacity. One company, BrightSource, dominates this space, with about 2.2 GW of projects contracted with two of California's IOUs (SCE and PG&E).
- Currently, there are no utility-scale linear Fresnel or parabolic dish systems contracted for development. The principal developers of both technologies, Ausra and Stirling Engine Systems, respectively, have met financial troubles that have disrupted their development pipelines. Ausra sold its linear Fresnel technology to French conglomerate Areva in 2010, and Stirling declared bankruptcy in September 2011.
- CPV technologies have made big gains in the development market, with nine projects totaling 471 MW currently under contract. This is notable for a technology that has only enjoyed a few years of commercialization.

Appendix A. Operating Utility-Scale Solar Plants

Table A-1. Operating Utility-Scale PV Plants

Plant	MW	Developer	Operating Year	Solar Tech	Location	PPA With
Aerojet	5	Solar Power Inc.	2010	c-Si	Sacramento, California	Aerojet Superfund Site
Agua Calliente (partial output)	30	First Solar	2012	CdTe	Yuma County, Arizona	MidAmerican Energy Holdings (Buffett)
Air Force Academy	5	SunPower	2011	c-Si	Colorado Springs, Colorado	
Alamogordo Solar Energy Center	5	First Solar	2011	CdTe	Deming, New Mexico	PNM
Alamosa Solar Plant	7	SunEdison	2007	c-Si	Hooper, Colorado	Xcel Energy
Avenal Solar Generating Facility	45	NRG	2011	a-Si	Avenal, California	PG&E
Blue Wing Solar Project	14	juwi	2010	CdTe	San Antonio, Texas	CPS Energy
Blythe Generating Facility	21	First Solar/NRG	2009	CdTe	Blythe, California	SCE
CalRenew-1	5	Meridian Energy	2010	a-Si	Mendota, California	PG&E
Cimarron I	30	First Solar	2010	CdTe	Cimarron, New Mexico	Tristate Generation and Transmission

Table A-1. (Continued)

Plant	MW	Developer	Operating Year	Solar Tech	Location	PPA With
Copper Mountain	48	Sempra Generation	2010	CdTe	Boulder City, Nevada	PG&E
Cotton Center Solar Plant	17	Solon	2011	c-Si	Gila Bend, Arizona	APS
Davidson County Solar	16	Sun Edison	2011	c-Si	Cotton Grove Township, North Carolina	Duke Energy
Deming Solar Energy Center	5	First Solar	2011	CdTe	Deming, New Mexico	PNM
Desoto Solar Energy	25	SunPower	2009	c-Si	Arcadia, Florida	FPL
Dover SUN Park	10	SunPower/LS Power	2011	c-Si	Dover, Deleware	Delmarva Power
El Dorado Energy Solar	10	First Solar/Sempra	2008	CdTe	Boulder City, Nevada	PG&E
Exelon City Solar	8	SunPower	2010	c-Si	Chicago, Illinois	Exelon Corp.
Five Points Station	15	Solon	2011	c-Si	Five Points, California	PG&E
FRV Webberville Plant	30	Fotowatio Renewable Ventures	2011	c-Si	Webberville, Texas	Austin Energy
Greater Sandhill I	18	SunPower	2010	c-Si	Mosca, Colorado	Xcel Energy
Hatch Solar Energy Center	5	NextEra/Amonix	2011	CPV	Hatch, New Mexico	El Paso Energy

Plant	MW	Developer	Operating Year	Solar Tech	Location	PPA With
Jacksonville Solar	12	juwi	2010	CdTe	Jacksonville, Florida	Jacksonville Electric Authority
Las Vegas Solar Energy Center	5	First Solar	2011	CdTe	Las Vegas, New Mexico	PNM
Long Island Solar Farm LLC	32	BP Solar	2011	c-Si	Upton, New York	Long Island Power Authority
Los Lunas Solar Energy Center	5	First Solar	2011	CdTe	Los Lunas, New Mexico	PNM
Mesquite Solar 1	42	Sempra Generation	2011	c-Si	Arlington, Arizona	PG&E
Nellis Solar	12	MMA Renewable Ventures	2007	c-Si	Nellis AFB, Nevada	Nellis AFB
Paloma Solar Plant	17	First Solar	2011	CdTe	Gila Bend, Arizona	APS
Paradise Solar Energy Center	5	NextEra Energy Resources	2010	c-Si	West Deptford, New Jersey	
Porterville	5	Cupertino Electric	2011	c-Si	Porterville, California	SCE
Road Runner Solar Electric Facility	20	NRG Energy	2011	CdTe	Santa Teresa, New Mexico	El Paso Electric
San Luis Valley Solar Ranch	30	Iberdrola	2011	c-Si	Alamosa County, Colorado	Xcel
SFPUC Sunset Reservoir	4	Recurrent Energy	2010	c-Si	San Francisco, California	Sunset Reservoir

Table A-1. (Continued)

Plant	MW	Developer	Operating Year	Solar Tech	Location	PPA With
Space Coast Solar Energy	10	SunPower	2010	c-Si	Titusville, Florida	FPL
SPS - Dollarhide	10	SunEdison	2011	c-Si	Dollarhide, New Mexico	SPS
SPS - Hopi	10	SundEdison	2011	c-Si	Hopi, New Mexico	SPS
SPS - Jal	10	SunEdison	2011	c-Si	Jal, New Mexico	SPS
Stanton Solar Farm	5	Regenesis Power	2011	c-Si	Orange County, Florida	PG&E
Stroud Solar Station	20	Cupertino Electric	2011	c-Si	Helm, California	PG&E
Westside Solar Station	15	Constellation Energy	2011	c-Si	Five Points, California	PG&E
Wyandot Solar	10	juwi	2010	CdTe	Upper Sandusky, Ohio	American Electric Power Co.
Total	**673**					

Appendix B. Utility-Scale Solar PV Projects under Development[28]

Table B-1. Distributed Utility-Scale PV Systems in Development (< 20 MW)

Plant	MW	Developer or Owner	Solar Tech	Location	PPA With
Austin Energy PV	5		PV	Texas	Austin Energy
Boron Solar Plant	15	Boron Solar, LLC	PV	California	SCE
Campbell Industrial Park Solar Project	5	Axio Power Inc.	PV	Hawaii	Hawaiian Electric Company, Inc.
Canton Landfill Solar Facility	5.6		PV	Massachusetts	NSTAR
Cascade Solar Project	18.5	Axio Power Inc.	PV	California	SCE
Catawba County Solar Project	5	Strata Solar LLC	PV	North Carolina	Duke Energy Carolinas, LLC
Celina Renewable Energy Center	5	SolarVision LLC	PV	Ohio	City of Celina
Central Antelope Dry Ranch B Solar Plant	5	Silverado Power LLC	PV	California	SCE
China Lake	14		PV	California	Naval Air Weapons Station
Concord Solar Farm	5	Shoe Show, Inc	PV	North Carolina	Duke Energy Corporation
Desert Green Solar Farm	6.5	Soitec USA Inc	CPV	California	SDG&E
Dos Rios Solar Project - I	10	SunEdison LLC	PV	Texas	CPS Energy
Dos Rios Solar Project - II	10	SunEdison LLC	PV	Texas	CPS Energy
East Lyme Solar Park Facility	5	GRE 214 East Lyme LLC	PV	Connecticut	United Illuminating Company
Foresight Solar PV Project	5	Foresight Renewables, LLC	PV	Arizona	Tucson Electric Power Company

Table B-1. (Continued)

Plant	MW	Developer or Owner	Solar Tech	Location	PPA With
Grundman Solar Project (Bruceville Road)	18	Constellation Energy Group, Inc.	PV	California	Sacramento Municipal Utility District
Heber Solar PV Project	10	Ormat Technologies, Inc.	PV	California	Imperial Irrigation District
Holiday Solar Array Facility	8.5	Clear Peak Energy, Inc	PV	California	SCE
Hyder Solar Plant	16		PV	Arizona	Arizona Public Service Company
Kalaeloa Renewable Energy Park (Solar)	5.9	Multi-Owned	PV	Hawaii	Hawaiian Electric Company, Inc.
Kalaeloa Solar II	5	SunPower Corporation	PV	Hawaii	Hawaiian Electric Company, Inc.
Kauai Solar Plant	6	Alexander & Baldwin, Incorporated	PV	Hawaii	Kauai Island Utility Cooperative
Keystone Solar Project	6	Community Energy Solar LLC	PV	Pennsylvania	Exelon Generation Company, LLC
Lakeland Linder Regional Airport Solar Project	5.5	SunEdison LLC	PV	Florida	City of Lakeland
Lancaster Dry Farm Ranch B Solar PV Plant	5	Silverado Power LLC	PV	California	SCE

Plant	MW	Developer or Owner	Solar Tech	Location	PPA With
Lancaster WAD B Solar PV Plant	5	Silverado Power LLC	PV	California	SCE
LanWest Solar Farm	6.5	Soitec USA Inc.	CPV	California	SDG&E
Littlerock Solar Power Generation Station	5	Amonix, Inc.	CPV	California	SCE
Lucerne Solar Power Generation Station	14	Amonix, Inc.	CPV	California	SCE
Luke Air Force Base Solar Project	15		PV	Arizona	Arizona Public Service Company
McDowell Technical Community College Solar Plant	5	McDowell Green Energy LLC	PV	North Carolina	Duke Energy Corporation
Middlesex Apple Orchard Complex	6.7	KDC Solar LLC	PV	New Jersey	County of Middlesex
Mitchell County Solar Project (Solar DD)	14	Solar Design & Development LLC	PV	Georgia	Georgia Power Company
Mohave County Solar Project	10	SOLON Corporation	PV	Arizona	UniSource Energy Services, Inc.
Mount St. Mary's PV Park	16	Constellation Energy Group, Inc.	CdTe	Maryland	State of Maryland
Murfreesboro Solar Project	6.5	Duke Energy Renewables	PV	North Carolina	North Carolina Electric Membership Corporation
Napoleon Solar I Project	10	BNB Napoleon Solar LLC	PV	Ohio	Campbell Soup Company
NJ Oak Solar Farm	10		PV	New Jersey	Atlantic City Electric

Table B-1. (Continued)

Plant	MW	Developer or Owner	Solar Tech	Location	PPA With
Orion Solar Project	12	Fotowatio Renewable Ventures Inc.	PV	California	PG&E
Palm Springs Solar PV Project-II	5	Multi-Owned	PV	California	SCE
PLP Kauai 1 Solar Thermal	10	Pacific Light & Power	PV	Hawaii	Kauai Island Utility Cooperative
Prescott Solar Facility	10	SunEdison LLC	PV	Arizona	Arizona Public Service Company
Queen Creek Solar Plant	19	PSEG Solar Source LLC	PV	Arizona	Salt River Project
RE Columbia 3 Solar Facility	10	Recurrent Energy Inc.	PV	California	SCE
REDCO Solar 1	5	Renewable Energy Development Corp.	PV	California	City of Needles
Rinehart Solar Farm Project - Stage I	10	BlueChip Energy, LLC	PV	Florida	Florida Power Corporation
Rio Grande Solar Plant	5	Recurrent Energy Inc.	PV	California	SCE
Ronkonkoma LIRR Station	5.5	enXco, Inc.	PV	New York	Long Island Power Authority
Sierra View Solar IV Plant	19	Juwi Solar Inc.	PV	California	SCE
Sierra View Solar V Plant	19	Juwi Solar Inc.	PV	California	SCE

Plant	MW	Developer or Owner	Solar Tech	Location	PPA With
Somers Solar Center Facility	5	HelioSage Energy, LLC	PV	Connecticut	United Illuminating Company
Somerset Solar Project	10	SunEdison LLC	PV	Texas	CPS Energy
South Robeson Solar Farm	5	Strata Solar LLC	PV	North Carolina	Carolina Power & Light Company
South Swan Solar PV Project	12	Amonix, Inc	CPV	Arizona	Tucson Electric Power Company
SunEdison NM Solar 4	10	SunEdison LLC	PV	New Mexico	Southwestern Public Service Company
SunEdison NM Solar 5	10	SunEdison LLC	PV	New Mexico	Southwestern Public Service Company
Tropico Solar PV Plant	14	Foresight Renewables, LLC	PV	California	SCE
UA Tech CTC Solar PV Project	5	CTC Solar	PV	Arizona	Tucson Electric Power Company
Victor Dry Farm Ranch A Solar PV Plant	5	Silverado Power LLC	PV	California	SCE
Victor Dry Farm Ranch B Solar Plant	5	Silverado Power LLC	PV	California	SCE
Vineland Solar One Expansion	12		PV	New Jersey	Vineland Municipal Electric Utility
West Tennessee Solar Farm	5	University of Tennessee	PV	Tennessee	Tennessee Valley Authority
Westford Solar Project	4.5	Multi-Owned	PV	Massachusetts	National Grid USA

Table B-1. (Continued)

Plant	MW	Developer or Owner	Solar Tech	Location	PPA With
Wharton Generating Station	10		PV	Texas	City of Houston
Westlands Solar Farms PV1	18	Westland Solar Farms, LLC	PV	California	PG&E
Total	**589.2**				

Table B-2. Mid-Sized PV Projects Currently in Development (20–50 MW)

Project	MW	Developer or Owner	Solar Tech	Location	PPA With
Adobe Solar Project	20	Fotowatio Renewable Ventures Inc.	PV	California	SCE
Alamosa Solar Generating Project	30	Cogentrix Energy, LLC	CPV	Colorado	Xcel
Alpaugh North Solar Plant	20	Solar Project Solutions, LLC	PV	California	PG&E
Antelope Solar Farm (Mojave Solar 4)	20	Fotowatio Renewable Ventures Inc.	PV	California	SCE
Antelope Solar Project - Tuusso	20	Multi-Owned	PV	California	SCE
Apex Solar Power Project	20	Fotowatio Renewable Ventures Inc.	PV	Nevada	NV Energy
Atwell Island Solar Plant	20	Solar Project Solutions, LLC	PV	California	PG&E
Avra Valley Solar Project	25	NRG Energy, Inc.	PV	Arizona	TEP

Project	MW	Developer or Owner	Solar Tech	Location	PPA With
Borrego Springs Project	26	NRG Solar LLC	PV	California	SDG&E
Central Antelope Dry Ranch C Solar Plant	20	Silverado Power LLC	PV	California	SCE
Chicago Rockford Solar Project	20	Multi-Owned	PV	Illinois	Ameren Illinois Company
Corcoran Solar Plant	20	Solar Project Solutions, LLC	PV	California	PG&E
Cygnus Solar Project	25	Fotowatio Renewable Ventures Inc.	PV	California	SCE
Del Sur Solar Power Plant	38	Beautiful Earth Group, LLC	PV	California	SCE
First Solar/PNM PPA	22	First Solar	CdTe	New Mexico	PNM
Florida Solar 1	25		PV	Florida	Tampa Electric
Holly Hill Grove 1	20	National Solar Power, LLC	PV	Florida	FPL
Holly Hill Grove 2	20	National Solar Power, LLC	PV	Florida	FPL
Holly Hill Grove 3	20	National Solar Power, LLC	PV	Florida	FPL
Holly Hill Grove 4	20	National Solar Power, LLC	PV	Florida	FPL
Holly Hill Grove 5	20	National Solar Power, LLC	PV	Florida	FPL
Holly Hill Grove 6	20	National Solar Power, LLC	PV	Florida	FPL
Holly Hill Grove 7	20	National Solar Power, LLC	PV	Florida	FPL
Holly Hill Grove 8	20	National Solar Power, LLC	PV	Florida	FPL
Holly Hill Grove 9	20	National Solar Power, LLC	PV	Florida	FPL
Holly Hill Grove 10	20	National Solar Power, LLC	PV	Florida	FPL
Illinois Solar	20		CdTe	Illinois	ComEd

Table B-2. (Continued)

Project	MW	Developer or Owner	Solar Tech	Location	PPA With
La Salle Solar Farm (Invenergy)	20	Invenergy LLC	CdTe	Illinois	Commonwealth Edison Company
LanEast Solar Farm	22	Soitec USA Inc	CPV	California	SDG&E
LSR Kramer South Solar	20	LSR Kramer South, LLC	PV	California	SCE
Marana Solar PV Project	35	Avalon Solar LLC	PV	Arizona	TEP
Maryland Solar Project	20	Maryland Solar LLC	PV	Maryland	FirstEnergy Solutions Corporation
McHenry Solar Farm	25	SunPower Corporation	PV	California	Modesto Irrigation District
Mililani Solar Park Project	20	Castle & Cooke Incorporation	PV	Hawaii	Hawaiian Electric Company, Inc.
Mojave Solar Project - Mojave	20	Fotowatio Renewable Ventures Inc.	PV	California	SCE
Mountain View Solar Plant	20	NextEra Energy Resources LLC	PV	Nevada	NV Energy
Murphy Flats Solar Project	20	Interconnect Solar Development, LLC	PV	Idaho	Idaho Power Co.
Nevada Solar Project	20	Fotowatio Renewable Ventures Inc.	PV	Nevada	NV Energy
Nicolis Solar PV Plant	20	Foresight Renewables, LLC	PV	California	SCE
Niland Solar Project	23	SunPeak Solar, LLC	PV	California	Imperial Irrigation District
North Edwards Solar	20	North Edwards Solar, LLC	PV	California	SCE

Project	MW	Developer or Owner	Solar Tech	Location	PPA With
North Lancaster Ranch Solar PV Plant	20	Silverado Power LLC	PV	California	SCE
RE Columbia 2 Solar Facility	20	Recurrent Energy Inc.	PV	California	SCE
RE Rosamond Solar Plant	20	Recurrent Energy Inc.	PV	California	SCE
San Luis Valley Solar Ranch Project	30	Iberdrola Renewables, Inc.	PV	Colorado	Xcel
Searchlight Solar I	20	American Capital Energy, Inc	PV	Nevada	NV Energy
Sierra Solar Greenworks Solar PV Plant	20	Silverado Power LLC	PV	California	SCE
Simon Solar Farm (Social Circle)	30	Private Investor- Steve Ivey	PV	Georgia	Georgia Power Company
Sorrento Eagle Dunes	40	BlueChip	PV	Florida	Progress Energy, Inc.
Spectrum Solar Power Plant	38	Fotowatio Renewable Ventures Inc.	PV	Nevada	NV Energy
TA High Desert Solar Plant	20	Multi-Owned	PV	California	SCE
Tierra Del Sol Solar Farm	45	Soitec USA Inc	CPV	California	SDG&E
Turning Point Solar	49.9	Agile Energy	PV	Ohio	Columbus Southern Power Co.
Victor Phelan Solar One Plant	20	Recurrent Energy Inc.	PV	California	SCE
Victorville Landfill Solar Project	21	Axio Power Inc.	PV	California	SCE

Table B-2. (Continued)

Project	MW	Developer or Owner	Solar Tech	Location	PPA With
Weldon Solar	20		PV	California	SCE
White River Solar Plant	20	Solar Project Solutions, LLC	PV	California	PG&E
Total	**1,329.9**				

Table B-3. Large-Sized PV Projects Currently in Development (> 50 MW)

Project	MW	Developer or Owner	Solar Tech	Location	PPA With
Agua Caliente	290	First Solar	CdTe	California	PG&E
Alhambra Solar Project	50	SolarGen2	PV	California	SDG&E
Alpaugh Solar	50	Solar Project Solutions, LLC	PV	California	PG&E
Alpine Solar	66	Alpine SunTower LLC	PV	California	PG&E
Amargosa Farm Road Solar Power Project	500	Multi-Owned	PV	Nevada	NV Energy
Amargosa North Solar Plant	150	Pacific Solar Investments Inc.	PV	Nevada	NV Energy
Antelope Valley Solar Ranch One	230	First Solar, Inc.	CdTe	California	PG&E
Arlington Valley Solar Energy Project II	125	LS Power Group	PV	California	SDG&E
AVSP I Solar Project	325	SunPower Corporation	PV	California	SCE
AVSP II Solar Project	276	SunPower Corporation	PV	California	SCE
California Valley Solar Ranch	250	NRG	PV	California	SCE
Catalina Solar Project	110	enXco Inc.	PV	California	SDG&E

Project	MW	Developer or Owner	Solar Tech	Location	PPA With
Centinela Solar Project	130	LS Power Group	PV	California	SDG&E
Copper Mountain Solar II	150	Sempra Generation	CdTe	California	PG&E
Desert Sunlight Project	550	First Solar Inc.	CdTe	California	PG&E
Estancia Solar Farm	50	Silverado Power	PV	New Mexico	PNM
FRV Regulus Solar Project	75	Fotowatio Renewable Ventures	PV	California	SCE
Gadsden Solar Farm	400	National Solar Power	PV	Florida	Progress Energy, Inc.
Hardee Solar Farm	200	National Solar Power	PV	Florida	Progress Energy, Inc.
Imperial Solar Energy Center South	130	Tenaska Solar Ventures	CdTe	California	SDG&E
Imperial Solar Energy Center West	250	Tenaska Solar Ventures	CPV	California	SDG&E
Mayflower Solar Project (SolarGen)	50	SolarGen2	PV	California	SDG&E
McCoy Solar Energy Project	750	NextEra Energy Resources LLC	PV	California	SCE
Mesquite Solar I	170	Sempra Generation	PV	Arizona	PG&E
Mount Signal Solar Project	50	US Solar Holdings	PV	California	SDG&E
National Solar Power Project	400	National Solar Power, LLC	PV	Florida	FPL
North Star Solar I Project	60	Multi-Owned	PV	California	PG&E
Pearl Harbor Solar Project	300	Sempra Generation	PV	Hawaii	Hawaiian Electric Industries, Inc.
Quinto Solar Project	110	SunPower Corporation	PV	California	SCE

Table B-3. (Continued)

Project	MW	Developer or Owner	Solar Tech	Location	PPA With
Rugged Solar Farm	80	Soitec USA Inc	CPV	California	SDG&E
Sacramento/Recurrent PPA	88	Recurrent Energy, Inc.	PV	California	Sacramento Municipal Utility District
Silver State North	140	NextLight Renewable Power, LLC	CdTe	Nevada	NV Energy
Silver State South	250	NextLight Renewable Power, LLC	CdTe	California	SCE
Solar Millennium Blythe Power Project (Photovoltaic)	1,000	Multi-Owned	PV	California	SCE
Solar Millennium Palen Power Project (CA Solar 10)	500	Multi-Owned	PV	California	SCE
Sonora Solar Project (SolarGen)	50	SolarGen2	PV	California	SDG&E
Sorrento Solar Farm	120	BlueChip Energy, LLC	PV	Florida	FPL
Stateline Solar Project	300	First Solar, Inc.	CdTe	California	SCE
Topaz Solar Farm	550	MidAmerican Energy	CdTe	California	PG&E
Wonder Valley Solar Farm	100	Sustainable Energy Capital Partners	PV	California	SCE
Total	**9,425**				

APPENDIX C. UTILITY-SCALE SOLAR AND POWER PURCHASE AGREEMENTS

Electric utilities, including investor-owned, municipal, and public, provide nearly all the electricity consumed by homes, businesses, and industries in the United States. Accordingly, their adoption of renewable energy systems is critical to the success of the technologies. In response to RPS requirements and customer demand for more renewable energy, electric utilities throughout the country have signed long-term contracts for the output of large-scale renewable energy generating facilities in recent years. These long-term contracts, known as power purchase agreements (PPAs), guarantee a steady and predictable revenue stream from the project, which lowers its risk profile and facilitates investment.

Due to stringent RPSs and other initiatives, California's three IOUs—PG&E, SCE, and SDG&E —lead in contracting with or developing solar projects. Figure 1 illustrates the significant gap between these utilities (particularly PG&E and SCE) in contracted solar megawatts relative to the other leading IOUs.

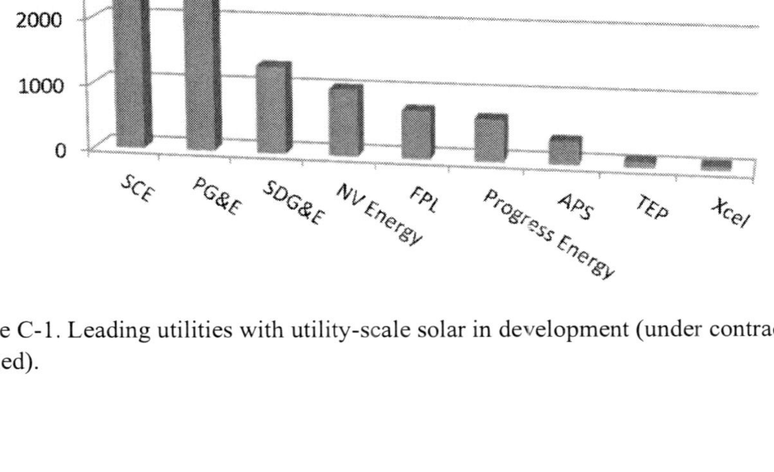

Figure C-1. Leading utilities with utility-scale solar in development (under contract or planned).

APPENDIX D. THE SOLAR RESOURCE

The solar resource, or quantity of sunlight, of the United States is significant. Total annual U.S. energy demand could theoretically be met with PV systems covering less than 1% of the nation's land area (Denholm and Margolis 2008). The southwest United States—comprising California, Arizona, Nevada, Utah, New Mexico, Colorado, and western Texas—has some of the best solar resources in the world. In fact, Arizona, California, Nevada, and New Mexico are estimated to have the solar resources to produce 4.1 million MW of electricity, almost four times the electric- generating capacity of the entire United States (DOE 2003; EIA 2008). The CSP and PV solar resource for the United States are shown in Figure D-1.

Despite abundant solar resources, significant economic barriers to widespread adoption remain. In particular, it is challenging to serve regions with modest solar resources, including arranging and underwriting the capital, building and maintaining the necessary transmission pathways, and assigning cost recovery through long-term contracts with utilities.

Furthermore, even in areas of high insolation (the measure of solar radiation striking a given area at a given time), finding suitable sites for large-scale solar installations can be difficult. Figure D-2 compares the availability of solar resource in the southern California and Nevada region—one of the best solar resource locations nationally—before and after considering various development factors. The left image depicts the regional resource with no "filters" or limitations to access. The right image depicts the same resource but with the following set of access limitations:

- Only premium solar resource (6.75 kWh/m^2/day and greater)
- No use of environmentally-sensitive or restricted land (e.g., excludes land controlled by the U.S. Department of Defense)
- No use of land with grades of 1% or greater.

As readily seen in the side-by-side graphics, applying the access filters greatly reduces the potential resource.

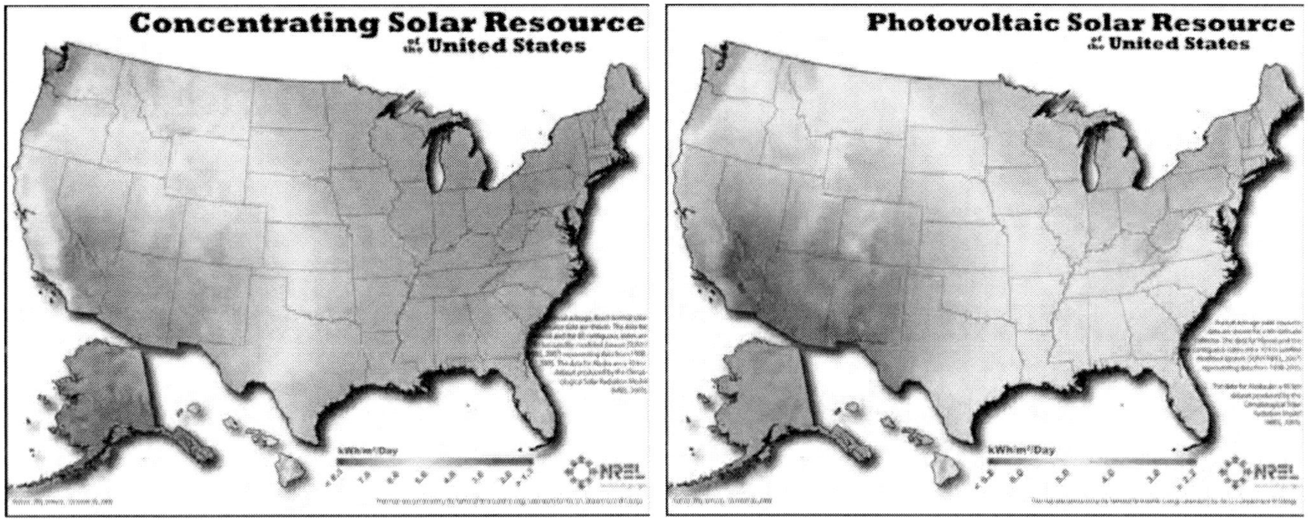

Source: NREL 2010c.

Figure D-1. Concentrating and PV solar resources in the United States.

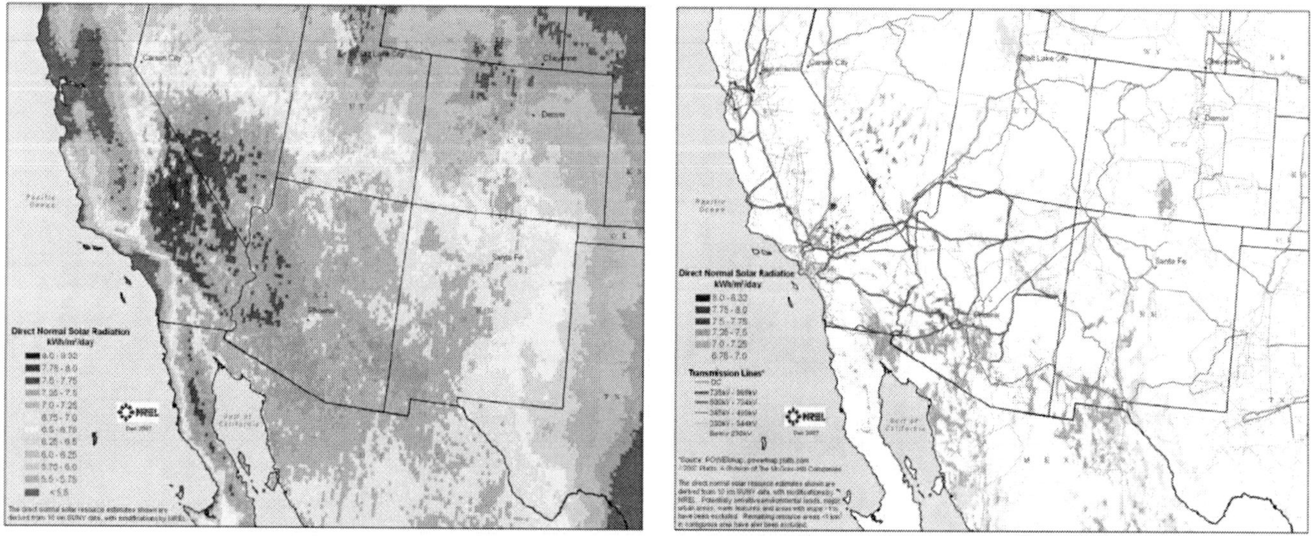

Figure D-2. Solar resource rear southern California without a filter (left) and with multiple filters applied (right).

REFERENCES

Acciona. (2010). "*Nevada Solar One*." http://www.acciona-na.com/About-Us/Our-Projects/U-S-/Nevada-Solar-One.aspx. Accessed July 2010.

American Capital Energy. (2009). "NV Energy and American Capital Energy Agree on Power Purchase Agreement for Solar PV from 20 MW DC Array in Southern Nevada." http://www.americancapitalenergy.com/nv-energy/. Accessed November 2, 2011.

Ausra. (2008). "Solar Power Company Ausra Launches First Solar Thermal Plant in California in Nearly 20 Years." http://www.ausra.com/news/releases/081023.html. Accessed November 2, 2011.

Australian National University. (February 2010). "Concentrating Solar Systems." http://solar-thermal.anu.edu.au/high-temperature/concentrating-solar-power-systems/. Accessed June 2010.

Baker, D. R. (2009). "OptiSolar Halts Manufacturing, Will Lay Off 200." *SF Gate*. http://articles.sfgate.com/2009-03-20/business/17215370_1_solar-power-plant-solar-panels-first- solar. Accessed November 2, 2011.

Baker, D. R. (2010). "*Engineering Giant Areva Buying Solar Firm*." *SF Gate*. http://articles.sfgate.com/2010-02-09/business/17872810_1_areva-solar-power-power-plants. Accessed November 2, 2011.

Barron, R. (2008). "Which Concentrating Solar Companies Are Doomed?" *Greentech Media*. http://www.greentechmedia.com/articles/read/which-concentrating-solar-companies-are- doomed-1394/. Accessed November 2, 2011.

Bates, S. P. (2000). "*Silicon Wafer Processing*." http://www.iisme.org/etp/silicon_wafer_processing.pdf. Accessed November 2, 2011.

Biello, D. (2009). "*How to Use Solar Energy at Night*." *Scientific American*. http://www.scientificamerican.com/article.cfm?id=how-to-use-solar-energy-at-night. Accessed November 2, 2011.

Bloomberg. (2009). "Solar Trust of America's U.S. Project Development Arm, Solar Millennium, LLC, to Employ Advanced Dry-Cooling Technology for Proposed Solar Thermal Power Plants in Nevada's Amargosa Valley." http://www.bloomberg.com/apps/news?pid=conewsstory&tkr=S2M:GR&sid=auzY6cG2CQJI. Accessed November 2, 2011.

BrightSource Energy. (2011a). "BrightSource Energy and Southern California Edison Add Energy Storage Capabilities to Power Purchase Agreements." http://www.brightsourceenergy.com/images/uploads/press_releases/BSE_SCE_PPA_Storage_11 2811_FINAL.pdf. Accessed February 16, 2012.

BrightSource Energy. (2011b). "Ivanpah Solar Power Complex – Ivanpah, California." http://www.brightsourceenergy.com/projects/ivanpah. Accessed June 2011.

Bureau of Land Management (BLM). (2010). "*Instruction Memorandum No. 2010-141*." http://www.blm.gov/wo/st/en/info/regulations/Instruction_Memos_and_Bulletins/national_instruction/2010/IM_2010-141.html. Accessed November 2, 2011.

BLM. (2011). "Solar Energy Plan of Development." http://www.blm.gov/pgdata/etc/medialib/blm/wo/MINERALS REALTY AND_RESOURCE_PROTECTION_/cost_recovery.Par.96285.File.dat/Solar_POD.pdf. Accessed November 2, 2011.

BusinessWire.com. (2009). "SunEdison and Xcel Energy Strike Deal for 50 MW (AC)/55 MW (DC) of Solar PV Energy in New Mexico." http://www.businesswire.com/portal/site/home/permalink/?ndmViewId=news_view&newsId=20091223005435&newsLang=en. Accessed November 2, 2011.

California Energy Commission (CEC). (2010a). "*City of Palmdale Hybrid Power Plant Project*." http://www.energy.ca.gov/sitingcases/palmdale/index.html. Accessed July 2010.

CEC. (2010b). "*Large Solar Energy Projects*." http://www.energy.ca.gov/siting/solar/index.html. Accessed November 2, 2011.

CEC. (August 2011). "*Energy Commission Approves Power Plant for Palmdale*." http://www.energy.ca.gov/releases/2011_releases/2011-08-10_Palmdale_pmpd_approved_nr.html. Accessed November 2, 2011.

California Environmental Protection Agency. (2008). "Water Quality Control Policy on the Use of Coastal and Estuarine Waters for Power Plant Cooling." http://www.energy.ca.gov/2008publications/SWRCB-1000-2008-001/SWRCB-1000-2008- 001.PDF. Accessed November 2, 2011.

Chakrabarty, G. (2009). "Xcel Selects SunPower to Build Solar Plant in Alamosa." *Denver Post*. http://www.denverpost.com/breakingnews/ci_12090370. Accessed November 2, 2011.

City of Victorville. (20 October 2008). "California Energy Commission Votes Unanimously to Approve Victorville 2 Hybrid Power Project." http://www.teamca.org/pdfs/2hybridpowerprojectapproval.pdf. Accessed November 2, 2011.

Clean Energy Authority. (2011a). "*Blythe Solar Project Converts from CSP to PV for its First Phase*." http://www.cleanenergyauthority.com/solar-energy-news/blythe-solar-project-converts-from-csp-to-pv-for-first-phase-082211/. Accessed November 2, 2011.

Clean Energy Authority. (2011b). "*Largest CPV Solar Plant in U.S. Completed in New Mexico*." http://www.cleanenergyauthority.com/solar-energy-news/largest-cpv-plant-in-us-completed-102611/. Accessed February 16, 2012.

Clean Energy Authority. (2012). "SunPower to Install First Commercial-Scale C7 Tracker System at ASU." http://www.cleanenergyauthority.com/solar-energy-news/sunpower-to-install-first-commercial-c7-tracker-011712/. Accessed February 16, 2012.

Deline, C., Marion, B., Granata, J. & Gonzalez, S. (January 2011). *A Performance and Economic Analysis of Distributed Power Electronics in Photovoltaic Systems*. Golden, CO: National Renewable Energy Laboratory. http://www.nrel.gov/docs/fy11osti/50003.pdf. Accessed February 16, 2012.

Delmarva Power. (2009). "*City of Dover, Delmarva Power, Partnering for 10 MW Solar Park*." http://www.delmarva.com/welcome/news/releases/archives/2009/article.aspx?cid=1209. Accessed November 2, 2011.

Denholm, P. & Margolis, R. (May 2008). "Impacts of Array Configuration on Land-Use Requirements for Large-Scale Photovoltaic Deployment in the United States." Presented at SOLAR 2008-American Solar Energy Society, San Diego, California.

U.S. Department of Energy (DOE). (15 September 2003). "*Why the Southwest U.S. Should Develop Its Solar Energy Resources*." http://www.solargenix.com/pdf/Southwest.pdf. Accessed November 2, 2011.

DOE. (December 2006). "*Energy Demands on Water Resources: Report to Congress on the Interdependency of Energy and Water*." http://www.sandia.gov/energy-water/docs/121-RptToCongress-EWwEIAcomments-FINAL.pdf. Accessed November 2, 2011.

DOE. (2008). "*Electricity Reliability Impacts of a Mandatory Cooling Tower Rule for Existing Steam Generation Units*." http://www.netl.doe.gov/energy- analyses/pubs/Cooling_Tower_Report.pdf. Accessed September 2011.

DOE. (2010a). "Concentrating Solar Power Commercial Application Study: Reducing Water Consumption of Concentrating Solar Power Electricity Generation Report to Congress." http://www.nrel.gov/csp/pdfs/csp_water_study.pdf. Accessed June 2010.

DOE. (2010b). "*Linear Concentrator Systems*." http://www1.eere.energy.gov/solar/linear_concentrator_rnd.html. Accessed August 2011.

DOE. (2010c). "*Solar Trough Power Plants*." http://www.osti.gov/accomplishments/documents/fullText/ACC0196.pdf. Accessed July 2010.

DOE. (2011a). "DOE Finalizes $150 Million Loan Guarantee to 1366 Technologies that Could Drive Down Manufacturing Costs and Make American Solar More Competitive. https://lpo.energy.gov/?p=5181. Accessed January 2012.

DOE. (2011b). "*Power Tower Systems for Concentrating Solar Power.*" http://www.eere.energy.gov/basics/renewable_energy/power_tower.html. Accessed October 2011.

DOE. (2011c). "*The Financing Force Behind America's Clean Energy Economy.*" Loan Programs Office. https://lpo.energy.gov/?page_id=45. Accessed October 2011.

DOE. (2011d). "*Concentrating Solar Power: Point Focus Reflector Technologies.*" Solar Energy Technologies Program. http://solareis.anl.gov/documents/docs/NREL_CSP_2.pdf. Accessed December 2011.

U.S. Department of the Interior. (2009). "Secretary Salazar, Senator Reid Announce 'Fast-Track' Initiatives for Solar Energy Development on Western Lands." http://www.blm.gov/wo/st/en/info/newsroom/2009/june/NR_0629_2009.html. Accessed November 2, 2011.

Energy Information Administration (EIA). (2008). "Existing Generating Units in the United States by State, Company and Plant, 2008." http://www.eia.doe.gov/cneaf/electricity/page/capacity/capacity.html. Accessed November 2, 2011.

Environmental Leader. (2010). "*Bell Building Concentrated Solar Plant with Thermal Storage System, First of its Kind.*" http://www.environmentalleader.com/2010/04/05/bell-building-concentrated-solar-plant-with-thermal-storage-system-first-of-its-kind/. Accessed November 2, 2011.

eSolar. (2009). "*eSolar Ushers in New Era of Solar Energy with Unveiling of Sierra Power Plant.*" http://esolar.com/news/press/2009_08_05. Accessed November 2, 2011.

eSolar. (2010). "*eSolar: Our Solution.*" http://www.esolar.com/our_solution/. Accessed June 2010.

Export-Import Bank of the United States. (2011). "*Ex-Im Bank Supports U.S. Renewable- Energy Jobs By Financing Solar-Power Projects In India.*" http://www.exim.gov/pressrelease.cfm/0C34ED47-DA59-908E-85498C3C62B91BB2/. Accessed November 2, 2011.

Fehrenbacher, K. (2009). "*Solar Firms are Slashing Jobs.*" http://earth2tech.com/2009/01/13/get-ready-for-the-solar-layoffs/. Accessed November 2, 2011.

Feo, E. & Tracy, S. (26 October 2009). "*Commercial Finance: The Dark Arts of Leverage, Tax Equity, Leases and More.*" Presented at Solar Power International: Anaheim, California.

First Solar. (2012). "*First Solar Sets Another World Record for CdTe Solar PV Efficiency.*" http://investor.firstsolar.com/releasedetail.cfm?ReleaseID=639 463. Accessed January 2012.

FPL. (2010). "Martin Next Generation Solar Energy Center." http://www.fpl.com/environment/solar/martin.shtml. Accessed November 2, 2011.

Garthwaite, J. (2010). "*First Solar Signs on PG&E for 300 MW of Desert Sunlight Project.*" http://earth2tech.com/2010/03/09/first-solar-signs-on-pge-for-300mw-of-desert-sunlight-project/. Accessed November 2, 2011.

Goodrich, A., James, T. & Woodhouse, M. (February 2012). *Residential, Commercial, and Utility- Scale Photovoltaic (PV) System Prices in the United States: Current Drivers and Cost-Reduction Opportunities.* NREL/TP-6A20-53347. Golden, CO: National Renewable Energy Laboratory. http://www.nrel.gov/docs/fy12osti/53347.pdf. Accessed March 12, 2012.

Green, M., Emory, K., Yoshihiro, H. & Warta, W. (January 2011). "Solar Cell Efficiency Tables (Version 37)." *Progress in Photovoltaics: Research and Applications*, (*19*,1), 84–92.

Green, S. (2009). "*NV Energy to Buy Power from Searchlight Solar Plant.*" http://www.lasvegassun.com/news/2009/jun/11/nv-energy-buy-power-searchlight-solar-plant/. Accessed November 2, 2011.

Hamberg, J. (2011). "Falling Silicon Prices Shakes up Solar Manufacturing Industry." *Down to Earth.* http://www.downtoearth.org.in/content/falling-silicon-prices-shakes-solar-manufacturing- industry. Accessed February 16, 2012.

Hermann, U., Geyer, M. & Kearney, D. (2002). "*Overview on Thermal Energy Systems.*" http://www.nrel.gov/csp/troughnet/pdfs/uh_storage_overview_ws030320.pdf. Accessed November 2, 2011.

IBM. (2011). "*Chevron Uses Solar Power to Produce More Oil.*" Smart Planet blog. http://www.smartplanet.com/blog/intelligent-energy/chevron-uses-solar-power-to-produce-more- oil/9302. Accessed January 2012.

Inland Energy. (2011). "*Reprint of Hybrid Plant Configuration.*" http://www.palmdale1.com/images/maingraphicconfig.jpg. Accessed October 2011.

Jaffe, M. (2009). "Xcel to Use Solar in Coal Plant." *Denver Post.* http://www.denverpost.com/breakingnews/ci_13219443?source=rss. Accessed November 2, 2011.

Kurtz, S. (June 2011). "*Opportunities and Challenges for Development of a Mature Concentrating Photovoltaic Power Industry*." Golden, CO: National Renewable Energy Laboratory. http://www.nrel.gov/docs/fy11osti/43208.pdf. Accessed July 2011.

LaMonica, M. (2009). "*Spain Plugs in Largest Solar-Tower Power Plant*." http://news.cnet.com/8301-11128_3-10228786-54.html. Accessed November 2, 2011.

Macknick, J., Newmark, R., Heath, G. & Hallett, K. (March 2011). "*A Review of Operational Water Consumption and Withdrawal Factors for Electricity Generating Technologies*." Golden, CO: National Renewable Energy Laboratory.

Martifer Renewables. (2010). "San Joaquin Solar 1 & 2 Hybrid Project, 08-AFC-12 -Notice of Withdrawal." Filed with CEC June 17, 2010. http://www.energy.ca.gov/sitingcases/sjsolar/documents/applicant/2010-06-17_Notice_of_Withdrawal_TN-57296.pdf. Accessed November 2, 2011.

Mendelsohn, M. (2010). "*Batting for Percentage: New Development Model May Lead to More RE Deployment*." http://financere.nrel.gov/finance/node/2841. Accessed November 2, 2011.

Murray, T. (2010). "*NRG Commits $300M to BrightSource Energy's Ivanpah Project*." http://greenenergyreporter.com/renewables/solar/nrg-commits-300m-brightsource-energys-392mw-facility/#more-11874. Accessed November 2, 2011.

NASDAQ QMS. (2006). "*SDG&E to Buy 120 Megawatts of Solar, Geothermal Energy*." http://www.globenewswire.com/newsroom/news.html?d=108622. Accessed November 2, 2011.

National Renewable Energy Laboratory (NREL). (2009). "*Genesis Solar Energy Project*." http://www.nrel.gov/csp/solarpaces/project_detail.cfm/projectID=54. Accessed November 2, 2011.

NREL. (2010a). "*Concentrating Solar Power Projects: Solana*." http://www.nrel.gov/csp/solarpaces/project_detail.cfm/projectID=23. Accessed June 2010.

NREL. (2010b). "*Sierra Suntower*." http://www.nrel.gov/csp/solarpaces/project_ detail.cfm/projectID=63. Accessed November 2, 2011.

NREL. (2010c). "*Solar Maps*." http://www.nrel.gov/gis/solar.html. Accessed November 2, 2011.

NREL. (2011a). "*CSP Trough Thermal Energy Storage Technology*." http://www.nrel.gov/csp/troughnet/thermal_energy_storage.html. Accessed November 2, 2011.

NREL. (2011b). "*PVWatts.*" http://www.nrel.gov/rredc/pvwatts/. Accessed November 2, 2011.

NextEra. (2010). "Solar Electric Generating Systems." http://www. next eraenergyresources.com/content/where/portfolio/pdf/segs.pdf. Accessed November 2011.

Nicola, S.; Doom, J. (2011). "*Solarhybrid Still Eager to Buy Solar Millennium's U.S. Projects.*" *Bloomberg.* http://www.bloomberg.com/ news/2011-12-22/solarhybrid-still-eager-to-buy-solar- millennium-s-u-s-projects.html. Accessed February 16, 2012.

NRG. (2010). "*Alpine Solar Project.*" http://nrgsolarenergy.com/pdf/alpine-factsheet-final.pdf. Accessed November 2, 2011.

NV Energy. (2010). "Solar Projects." http://www.nvenergy.com/renewables environment/renewables/solar.cfm. Accessed August 2010.

Outland Renewable Energy. (2010). "*Allco Renewable Energy to Develop $45,000,000 Solar Project in Rhode Island.*" http://outlandenergy.com/ allco-renewable-energy-to-develop- 45000000-solar-project-in-rhode-island/. Accessed November 2, 2011.

Pankratz, H. (2011). "Alamosa Solar Facility, Expected to be the Largest of its Kind, Receives $90.6 M Loan Guarantee from Feds." *The Denver Post.* http://www.denverpost.com/breakingnews/ci_18860250.Accessed February 16, 2012.

PG&E. (2008). "*PG&E Signs Historic 800 MW Photovoltaic Solar Power Agreements With Optisolar and Sunpower.*" http://www.pge.com/about/ news/mediarelations/newsreleases/q3_2008/080814.shtml. Accessed November 2, 2011.

PG&E. (2009). "*Contract between PG&E and Rice Solar Energy, LLC (A Subsidiary of Solar Reserve, LLC) for the Procurement of Renewable Energy Resources.*" http://www.pge.com/nots/rates/tariffs/tm2/pdf/ELEC_ 3581-E.pdf. Accessed October 2011.

PG&E. (2010). "*PG&E and BrightSource Sign Record Solar Power Deal.*" www.brightsourceenergy.com/images/uploads/press_releases/PGE_Bright Source_051309.pdf. Accessed November 2, 2011.

Pitz-Paal. (2009). "*How the Sun Gets Into the Power Plant.*" http://www. docstoc.com/docs/26711733/Concentrating-Solar-Power-(CSP). Accessed April 2011.

Price, T. (November 2009). "Molten Salt: The Magic Ingredient." *CSP Today.* http://social.csptoday.com/industry-insight/molten-salt-magic-ingredient. Accessed November 2, 2011.

Prior, B. (December 2010). "Abengoa Closes $1.45B DOE Loan Guarantee for Solana CSP Plant." *Greentech Media*. http://www.greentechmedia.com/articles/read/abengoa-closes-1.45b- doe-lo/. Accessed November 2, 2011.

PSE&G. (2010). "*PSE&G Solar 4 All™ Program.*" http://www.pseg.com/family/pseandg/solar4all/index.jsp. Accessed October 2011.

PV Magazine. (2011). "*Solarhybrid's Solar Millennium PV Pipeline Takeover Postponed.*" http://www.pv-magazine.com/news/details/beitrag/solarhybrids-solar-millennium-pv-pipeline-takeover-postponed_100005314/. Accessed February 17, 2012.

PV Magazine. (2012). "*Solarhybrid Files for Bankruptcy.*" http://www.pv-magazine.com/news/details/beitrag/solarhybrid-files-for-bankruptcy_100006188/#axzz1pjo8dKfm. Accessed April 3, 2012.

Recurrent Energy. (2010). "*The San Francisco Sunset Reservoir Solar Project.*" http://www.recurrentenergy.com/projects/sfsunset.php. Accessed November 2, 2011.

Renewable Energy Research Laboratory. (2011). "*Wind Power: Capacity Factor, Intermittency, and What Happens When the Wind Doesn't Blow?*" University of Massachusetts at Amherst. http://www.umass.edu/windenergy/publications/published/communityWindFactSheets/RERL_Fact_Sheet_2a_Capacity_Factor.pdf. Accessed September 2011.

Renewable Energy World. (2009c). "*LIPA Approves Largest Solar Energy Project in New York State.*" http://www.renewableenergyworld.com/rea/news/article/2009/12/lipa-approves-largest-solar-energy-project-in-new-york-state. Accessed November 2, 2011.

Reuters. (2009). "*Utility PG&E, SolarReserve Sign Power Contract.*" http://www.reuters.com/article/idUSN2219585620091223. Accessed November 2, 2011.

Richter, C., Short, R. & Teske, S. (2009). "*Concentrating Solar Power Global Outlook 09: Why Renewable Energy is Hot.*" ESTELA, Green Peace International, and Solar Paces. Amsterdam, Netherlands.

Russell, M. C. (2010). "*The Promise of Reliable Inverters for PV Systems: The Microinverter Solution.*" http://www.greentechmedia.com/articles/read/the-promise-of-reliable-inverters-for-pv-systems-the-micro-inverter-solutio/. Accessed November 2, 2011.

San Diego Gas and Electric (SDG&E). (2008). "SDG&E Solar Energy Project." www.sdge.com/regulatory/documents/a-08-07-017/CoverSheets ExecSummary.doc. Accessed November 2, 2011.

Solana Solar. (2009). "*Solana Generating Station Project.*" http://solanasolar.com/misc/Solana.pdf. Accessed November 2, 2011.

SolarBridge. (2010). "*SolarBridge Microinverter Solution Enables AC Modules with 25-Year Warranty, Higher Energy Harvest at Lower Cost.*" http://solarbridgetech.com/solarbridge- microinverter-solution-enables-ac-modules-with-25-year-warranty-higher-energy-harvest-at- lower-cost/. Accessed November 2, 2011.

Solarbuzz. (2010). "*Technologies.*" http://www.solarbuzz.com/going-solar/understanding/technologies. Accessed January 2012.

Solar Energy Industries Association (SEIA). (2011a). "*Utility-Scale Solar Projects in the United States Operating, Under Construction, or Under Development.*" http://www.seia.org/galleries/pdf/Major%20Solar%20Projects.pdf. Accessed January 2012.

SEIA. (2011b). "*U.S. Solar Market Insight, 2010 Year in Review: Executive Summary.*" http://www.seia.org/galleries/pdf/SMI-YIR-2010-ES.pdf. Accessed January 2012.

SEIA. (2012). "*U.S. Solar Market Insight, 2011 Year in Review: Executive Summary.*" http://www.seia.org/galleries/pdf/SMI-YIR-2011-ES.pdf. Accessed March 2012.

Solar Foundation. (October 2010). "*National Solar Jobs Census.*" http://www.thesolarfoundation.org/sites/thesolarfoundation.org/files/Final%20TSF%20National%20Solar%20Jobs%20Census%202010%20Web%20Version.pdf. Accessed November 2, 2011.

Solar Millennium. (2010). "*The Parabolic Trough Power Plants Andasol 1 to 3.*" http://www.solarmillennium.de/upload/Download/Technologie/eng/Andasol1-3engl.pdf. Accessed November 2, 2011.

SolarPaces. (2010). "*Solar Dish Engine.*" http://www.solarpaces.org/CSP_Technology/docs/solar_dish.pdf. Accessed June 2010.

Solar Power Partners. (2010). "Ivanpah Solar Electric Generating System (ISEGS) (07-AFC-5) Biological Mitigation Proposal, Submitted to the Bureau of Land Management and the California Energy Commission." http://www.energy.ca.gov/sitingcases/ivanpah/documents/applicant/2010-02-11_Biological_Mitigation_Proposal_TN_55374.pdf. Accessed November 2, 2011.

SolarReserve. (2011). http://www.solar-reserve.com/aboutUs.html. Accessed November 2, 2011. *Solar Thermal Magazine.* (2010). "Solar CSP Trough CSP Plant with Bell Energy Storage Technology (BEST) System." (2010). http://www.solarthermalmagazine.com/2009/09/17/solar-parabolic-trough-csp-plant-with-bell-energy-storage-technology-best-system/. Accessed June 2010.

Southern California Edison (SCE). (2010). "Solar Rooftop Program." http://www.sce.com/solarleadership/solar-rooftop-program/general-faq.htm. Accessed November 2, 2011.

Stoddard, L. (13 February 2008). "Evaluating Central Station Solar Technologies." EUCI Course on Centralized Solar for Electric Generation. Denver, CO: Black & Veatch.

SunPower. (2010). "*E19/425 Solar Panel: Maximum Efficiency and Performance*." http://us.sunpowercorp.com/cs/BlobServer?blobkey=id&blobwhere=1300258501532&blobheadername2=Content-Disposition&blobheadername1=Content-Type&blobheadervalue2=inline%3B+filename%3DSP425%2BPanel%2BSpec%2BSheet%2Bfor%2BDownload%252C1.pdf&blobheadervalue1=application%2Fpdf&blobcol=urldata&blobtabl e=MungoBlobs. Accessed January 2012.

SustainableBusiness.com. (2010). "*SunPower, PG&E Expand Power Plant Plans to 250 MW*." http://www.sustainablebusiness.com/index.cfm/go/news.display/id/20226. Accessed November 2, 2011.

Thurston, C. W. (2010). "*Areva Boosts Solar Supersteam Parameters in Bakersfield*." http://www.renewableenergyworld.com/rea/news/article/2010/07/areva-boosts-solar-supersteam-parameters-in-bakersfield?cmpid=rss. Accessed November 2, 2011.

Tribal Energy and Environment Information Clearinghouse. (2010). "*Solar Energy System Descriptions*." http://teeic.anl.gov/er/solar/restech/desc/index.cfm. Accessed June 2010.

University of Tennessee. (2010). "*Volunteer State Solar Initiative Helps Tennessee Lead in Renewable Energy Sector*." http://www.tennessee.edu/media/releases/071210_solar.html. Accessed November 2, 2011.

Wang, U. (2009). "*First Solar Inks 550 MW Deals with SoCal Edison*." http://www.greentechmedia.com/articles/read/first-solar-to-build-550mw-for-socal-edison/. Accessed November 2, 2011.

Wang, U. (2010). "*Energy Conversion Devices to Lay Off 140 in Michigan, Shift Some Manufacturing to Mexico*." http://www.dailyfinance.com/story/investing/energy-conversion-devices-to-lay-off-140-in-michigan-shift-some/19589777/. Accessed November 2, 2011.

Wesoff, E. (2010). "*SCE Signs Contracts for 240 MW of PV and CPV*." http://www.greentechmedia.com/articles/read/SCE-Signs-Contracts-For-225MW-of-PV-and- CPV/. Accessed November 2, 2011.

Wesoff, E. (2011). "*BrightSource's 'Head Start' for Desert Tortoises.*" http://www.greentechmedia.com/articles/read/brightsource-energys-head-start-program-for- desert-tortoises/. Accessed November 2, 2011.

Wesoff, E. & Prior, B. (2011). "*German Solar Shakeout: Solar Millennium Insolvency.*" http://www.greentechmedia.com/articles/read/German-Solar-Shakeout-Solar-Millennium- Insolvency/. Accessed February 17, 2012.

Wisconsin Public Service Commission. (2009) "*Electric Transmission Lines.*" http://psc.wi.gov/thelibrary/publications/electric/electric09.pdf. Accessed June 2010.

Vo, D. (2010). "*Duke Energy Buys 14 Mw Blue Wing Solar Project.*" http://solar.coolerplanet.com/News/1261001-duke-energy-buys-14mw-blue-wing-solar- project.aspx. Accessed November 2, 2011.

End Notes

[1] This report categorizes CPV as a PV technology, though some analysts group it under CSP.

[2] This database was corroborated by similar databases maintained by the Solar Energy Industries Association (SEIA 2011a) and SNL Financial.

[3] For this paper, "utility-scale" is defined as projects 5 MW or larger. These projects were either publicly announced and hold a long-term power purchase agreement or were announced directly by a utility. Public announcements are made via press releases.

[4] In the energy industry, some, if not many, planned projects will not reach completion. Therefore, we assume this figure to be greater than what will be delivered by the current pipeline of projects.

[5] Copper-indium-gallium-selenide (CIGS) is perhaps the more common version of this thin-film technology. Solar Frontier, the sole supplier of CIS/CIGS thin-film modules to the utility-scale market (as of January 2012), does not use gallium in their semi-conductor blend.

[6] Solar Millennium also filed for insolvency in December 2011 (Wesoff and Prior 2011).

[7] RPS policies are essentially mandated quotas for renewable energy generation as a proportion of total electricity production.

[8] For this paper, "utility-scale" is defined as any solar electric system with a capacity of 5 MW and above. Such utility-scale installations can deploy solar technologies far faster than traditional "behind-the-meter" projects designed to offset retail load. These systems employ significant economies of scale during construction, operation, and financial capital attraction, which can reduce the delivered cost of power.

[9] Beyond SEGS and Nevada Solar One, applications exist in Israel, Algeria, and Spain.

[10] PPAs are contracts between power producers and power purchasers for the long-term sale of electricity. See Appendix B for more information.

[11] In March, 2012, solarhybrid began its own insolvency proceeding due to concerns of illiquidity (i.e., not enough cash to pay bills) (PV Magazine 2012).

[12] In March 2011, the Export-Import Bank of the United States supplied a direct loan of $30 million to develop a 10 MW solar dish project in Rajasthan, India. U.S.-based dish manufacturer Infina Corporation will supply the modules for this project (Export-Import Bank 2011).

[13] For purposes of this report, in NREL's database projects are designated as hybrid if at least 50% of the energy is expected to be derived from fossil fuels. Many CSP systems utilize a

small quantity of fossil fuel but are not classified as hybrid systems. For example, the BrightSource Ivanpah project will utilize a small auxiliary boiler, which is expected to provide 2% of its output.

[14] Capacity factor represents the delivered energy production divided by the theoretic energy production if the plant operated at full output all the time.

[15] Because the generator size will be smaller than actual capacity after the application of storage, these 250 MW are a "net" figure.

[16] According to a BrightSource press release, "Under the original power purchase agreements with Southern California Edison, BrightSource would provide approximately four million megawatt-hours of electricity annually across seven power plants. Due to higher efficiencies and capacity factors associated with energy storage, the new set of agreements will provide approximately the same amount of energy annually but with one less plant, reducing the land impacts of delivering this energy and avoiding transactional costs that ultimately impact California's ratepayers" (BrightSource Energy 2011a, p. 2).

[17] Solar panel efficiency is the ratio of electric power produced by a PV module to the power of the sunlight striking the module.

[18] This includes installations at the residential, commercial, and utility scale. If only utility-scale installations (as defined by this report) are considered, then the figure is approximately 58%, according to NREL's internal databases.

[19] The installation of micro-inverters on each module can recover 10%–30% of annual performance loss due to shading. This has proven a more efficient approach to DC/AC conversion than the installation of one central inverter (Deline et al. 2011).

[20] SunPower's E19/425 solar panel has a reported conversion efficiency of 19.7%, one of the highest in the PV market today (SunPower 2010).

[21] For more information, visit www.recurrentenergy.com/solar/.

[22] Further, large customers pay for power primarily through demand charges tied to their peak demand during the month (and often with a "ratchet" mechanism tied to the peak demand in the prior 12-month period). Accordingly, due to the variable output of a solar field, utility charges may not decline proportionately with solar production, although the energy component of the utility bill should decline proportionately with solar production.

[23] Regulatory decisions to limit rate increases can negatively impact bond ratings and the utility's ability to counter-sign renewable energy contracts or develop renewable energy projects. The importance of a utility's creditworthiness is discussed in greater length in the second report, *Finance and Government Initiatives* (Mendelsohn et al. 2011).

[24] CIS is a variant of copper-indium-gallium-diselenide, often referred to as CIGS.

[25] Another CPV manufacturer, Solfocus, has installations in the United States, but none are utility scale as defined in this report.

[26] Low and medium concentrating systems may soon be competitive as well. For example, in late 2011 SunPower entered the CPV market with the rollout of their C7 Tracker system. This technology uses parabolic mirrors instead of lenses to concentrate sunlight, and monocrystalline silicon cells instead of multi-junction cells. The company has not yet contracted this technology for any utility-scale projects, though a 1 MW installation is planned at Arizona State University's Polytechnic campus (Clean Energy Authority 2012).

[27] There are larger CPV projects in development, but none are expected to be completed as soon as the Alamosa Project.

[28] Denoting "PV" in the "Tech" column indicates that the panels to be selected will most likely (but not certainly) be c-Si.

In: Utility-Scale Solar Installations
Editor: Miles B. Simmons
ISBN: 978-1-62948-999-5
© 2014 Nova Science Publishers, Inc.

Chapter 2

FEDERAL AND STATE STRUCTURES TO SUPPORT FINANCING UTILITY-SCALE SOLAR PROJECTS AND THE BUSINESS MODELS DESIGNED TO UTILIZE THEM*

Michael Mendelsohn and Claire Kreycik

LIST OF ACRONYMS

ACP	alternative compliance payment
AMT	alternative minimum tax
CSP	concentrating solar power
D.C.	District of Columbia
DG	distributed generation
DOE	Department of Energy
EESA	Emergency Economic Stabilization Act of 2008
FIPP	Financial Institute Partnership Program
IOU	investor-owned utility
IPP	independent power producer
IRS	Internal Revenue Service
ISO	independent system operator

* This is an edited, reformatted and augmented version of the U.S. Department of Energy, Office of Energy Efficiency & Renewable Energy, NREL/TP-6A20-48685, dated April, 2012.

ITC	investment tax credit
LADWP	Los Angeles Department of Water and Power
LSE	load-serving entity
MACRS	Modified Accelerated Cost Recovery System
MW	megawatt
MWh	megawatt-hour
NREL	National Renewable Energy Laboratory
PG&E	Pacific Gas and Electric
PPA	power purchase agreement
PSE&G	Public Service Electric & Gas Company
PTC	production tax credit
PV	photovoltaic
REC	renewable energy certificate
Recovery Act	American Recovery and Reinvestment Act
REFTI	Renewable Energy Finance Tracking Initiative
RPG	renewable portfolio goal
RPS	renewable portfolio standard
SCPPA	Southern California Public Power Authority
SDG&E	San Diego Gas and Electric
SREC	solar renewable energy certificate

EXECUTIVE SUMMARY

Over the past two decades, state and federal policymakers have enacted numerous policies to support new renewable energy development. The following report evaluates those policies based on industry literature, publicly available data, and questionnaires conducted by the National Renewable Energy Laboratory (NREL). As such, the report is relevant to policymakers who want to enable project financing via effective legislation, to regulators interested in supporting new project development, and to new investors interested in entering the renewable energy project finance space.

Three policy types have been critical to the widespread deployment of utility-scale solar facilities:

1. Federal tax benefits comprising the 30% investment tax credit (ITC) and accelerated depreciation schedules, and the Section 1603 cash grant program available in lieu of the ITC

2. The Department of Energy (DOE) loan guarantee program, which has provided guarantees and access to low-cost financing, enabling larger solar energy projects than ever deployed before
3. State renewable portfolio standards (RPS), enacted by 29 states and the District of Columbia (D.C.), which have created a necessary demand for renewable energy, and in particular, have led to the signing of long-term contracts critical to financing large-scale solar facilities.

As of February 21, 2012, the Section 1603 (also known as the "cash grant") program had awarded $10.76 billion in grants to roughly 23,000 projects (Department of Treasury 2012). The cash grant program expired in 2011 but will award projects that started construction in 2009, 2010, or 2011.

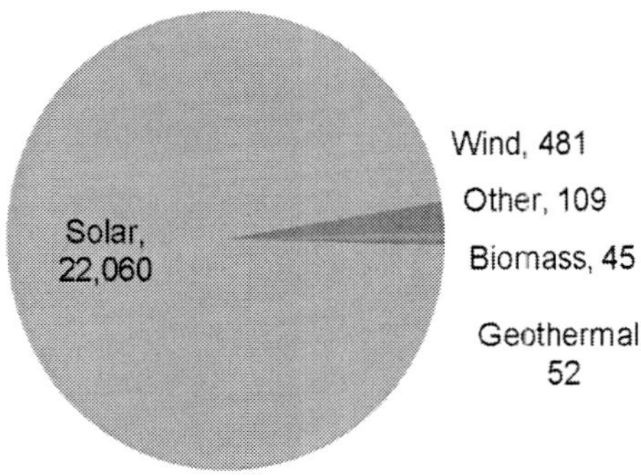

Source: Department of Treasury 2012.

Figure ES-1. Grants awarded by technology.

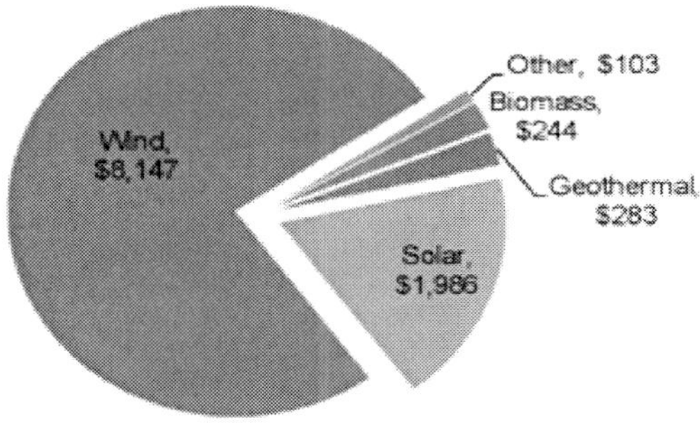

Source: Department of Treasury 2012.

Figure ES-2. Funding awarded by technology ($millions).

Utility-scale solar projects in particular benefitted significantly from the loan guarantee program. The loan guarantee program provides access to capital at a lower cost and larger quantity than what private markets would provide without the program. Through January 2012, solar projects have been supported solely under Section 1705 of the loan guarantee program, through which over $12 billion in loans have been made to 3,500 MW of solar projects. To support the solar project loans, an estimated $1.4 billion in credit subsidy costs have been paid under the American Recovery and Reinvestment Act, as shown in Table ES-1.

The credit subsidy is essentially a fund set aside by the DOE to cover the costs associated with specific project failures. As the loan guarantee program required an average credit subsidy of 11.7% per project, every dollar the DOE spends in credit subsidy will support $8.55 in loans. Assuming an average financial structure of 60% debt and 40% equity (actual equity investments are not known), each credit subsidy dollar invested can actually be extended to support $14.25 in solar project development.

Of note, the private bond market—without any loan guarantees—has rebounded from the financial crisis. In February 2012, to finance the Topaz purchase, MidAmerican Energy successfully offered $850 million of private placement bonds and is considering a second bond offering to raise an additional $430 million.

State RPSs create critical demand for large-scale generation projects, driving investor- owned and public utilities to sign long-term purchase power agreement (PPA) contracts with solar project developers. These PPAs are fundamental to project financing as debt and equity investors will not risk capital without the surety of revenue from a creditworthy counter-party to the contract.

However, the solar industry is capable of far greater deployment than currently required under the collective RPSs presently enacted. The Lawrence Berkeley National Laboratory recently estimated that nine states and D.C. have specific solar renewable energy certificate requirements and will require incremental solar capacity of 6,729 MW by 2025 (Barbose 2011). That represents roughly 480 MW of solar capacity installed per year, less than half the total domestic installations in 2010. Even in the most aggressive state—New Jersey—current solar project deployment meets the build-out requirements for 2025.

Table ES-1. Loan Guarantees for Solar Generation Projects

Technology	Loan Amount ($billions)	Estimated Credit Subsidy at 11.7% Rate ($billions)
Solar Photovoltaic (PV)	6.14	0.72
CSP	5.86	0.69
Solar Manufacturing	1.28	0.15
Wind	1.70	0.20
Other Technologies	1.19	0.14
Total	**16.16**	**1.88**
Solar Electric Generation Only	*12.00*	*1.40*

Although the array of federal and state support mechanisms greatly improved the ability of solar energy projects to achieve financing, complex financial structures were still required to utilize the depreciation benefits and access investment capital. The advanced financial structures described in this report include:

- Single owner finance, which includes utility balance sheet financing and developer balance sheet financing
- Lease structures generally referred to as sale-leaseback and inverted leases
- Partnership flips
- Utility prepay (which has been solely utilized by the wind industry only modestly to date).

This report is NREL's second study in a three-part series on utility-scale solar in the United States. In the first study (Mendelsohn et al. 2012b), NREL found that, as of January 2012, approximately 16,000 MW of utility-scale solar projects (defined as 5 MW or larger) are in various stages of advanced development and hold a contract with an investor-owned or public utility. If all these projects come to fruition, the amount of solar in the United States will increase approximately six-fold. The third study (Mendelsohn et al. 2012a) is a quantitative analysis on the cost impact of specific financial structures and terms as they relate to utility-scale solar facilities.

1. INTRODUCTION

Utility-scale solar projects have grown rapidly in number and size over the last few years, driven in part by strong renewable portfolio standards (RPS) and federal incentives designed to stimulate investment in renewable energy technologies. This report provides an overview of such policies, as well as the project financial structures they enable. For the purposes of this report, utility-scale solar plants are defined as generators 5 MW or larger that have a signed power purchase agreement (PPA) with an electric utility (or are being developed by an electric utility).

This report is the second in a three-part series regarding utility-scale solar markets, policies, and financing practices. The first part, *Utility-Scale Solar Concentrating Solar Power and Photovoltaics Projects: A Technology and Market Overview* (Mendelsohn et al. 2012b), describes solar power generation technologies and characterizes the market landscape and projects in the pipeline. The third part, *The Impact of Financial Structure on the Cost of Solar Energy* (Mendelsohn et al. 2012a), provides a quantitative analysis of financial structure and term and insight from renewable energy financial professionals.

The first report, providing technology and market overview, finds that as of January 2012, slightly over 16,000 MW of utility-scale solar projects are in various stages of advanced development and hold contracts with an investor-owned or public utility. Photovoltaic (PV) technologies account for 72% of this capacity, and concentrating solar power (CSP) technologies account for about 26%. The remaining 2% is comprised of projects that utilize unconventional solar technologies. Utility-scale installations represent the fastest growing segment of the solar industry. Third quarter (Q3) 2011 installations of over 200 MW were about as large as total 2010 utility-scale installations (SEIA 2011).

2. POLICIES SUPPORTING UTILITY-SCALE SOLAR DEVELOPMENT

Federal and state policies provide crucial support to the solar industry in the form of tax incentives, grants, loan guarantees, renewable energy standards, and production incentives.

2.1. Federal Economic Recovery Policies

The Energy Policy Act of 2005 increased the investment tax credit (ITC) for solar energy systems from 10% to 30% of eligible solar property (SEIA 2011). The ITC reduces the overall tax liability for individuals or businesses that make investments in solar energy generation technology.

The Emergency Economic Stabilization Act of 2008 (EESA) extended the 30% ITC for solar and other qualifying renewable technologies through December 31, 2016 (Duval & Stachlenfeld 2008).[1] EESA expanded the ITC use by making it available to public utilities (investor-owned utilities). Prior to EESA, public utilities were not eligible for ITC benefits.

The American Recovery and Reinvestment Act (Recovery Act), passed in February 2009, further expanded the availability and usability of various tax credits, depreciation opportunities, loan guarantees, and other mechanisms designed to encourage private investment in renewable energy and energy efficiency projects (Chadbourne & Parke 2010b). Specifically, the Recovery Act enabled ITC-eligible projects to receive a grant of equivalent value in lieu of the tax credit. The U.S. Department of the Treasury grant program is described in Section 2.1.2.

2.1.1. Investment Tax Credit Provisions

The ITC reduces federal income taxes for qualified owners based on capital investment in renewable energy projects. The ITC is realized in the year in which the solar project begins commercial operations, but Internal Revenue Service (IRS) regulations require the asset be retained for a five-year period during which the asset vests to the owner. As dictated by the "clawback" provision, the IRS will recapture any unvested portion of the credit if the project owner sells the project before the end of the fifth year of commercial operations (Bolinger 2009a).[2]

In addition to extending the 30% ITC for solar and other qualifying renewable technologies through December 31, 2016, EESA allows the ITC to offset both regular and alternative minimum tax (AMT) and permits utilities to claim the ITC when directly investing in solar facilities (although "normalization" rules complicate the benefit to utilities and their ratepayers[3]).

"Passive activity loss rules," also known as "passive income" rules, also come into play with the ITC. These rules were enacted as part of the Tax Reform Act of 1986 to prevent the use of losses from activities in which an investor did not actively participate to offset that investor's wage, interest, and dividend income. Passive activities involve the conduct of a trade or business in which the taxpayer does not materially participate, such as rental income (Tracy 2010).[4] This limitation is known to constrain the availability of tax equity.

2.1.2. Investment Tax Credit and Treasury Grants

Under Section 1603 of the Recovery Act, qualifying commercial renewable energy projects can accept a cash payment in lieu of either the ITC or production tax credit (PTC). The payment is generally referred to, albeit incorrectly, as a cash grant, and the program as the Treasury grant program.[5] Via the program, the Treasury pays grants equal to 30% of the cost of solar property placed in service (Bolinger et al. 2009). The program was originally scheduled to terminate on December 31, 2010, but Congress extended it through December 31, 2011. Property that is not placed in service prior to December 31, 2011, can qualify for the grant program as long as construction has commenced prior to that date. In order to be considered under construction, a project must begin "physical work of a significant nature" and conduct continuous construction (excluding preliminary work like planning, designing, and preparing land) or meet a "safe harbor" requirement, whereby 5% of total costs are incurred by the end of 2011. In either case, the solar facility must be placed in service before January 1, 2017 (Chadbourne & Parke 2010a; Hastings 2010).

Importantly, the 1603 Treasury grant is not a reduction in taxes and, thus, does not require taxable income to offset. However, only entities that pay taxes (and thus are eligible for the ITC) are eligible to receive the Treasury grant, negating the opportunity for municipal entities and non-profits to directly participate in the program.

According to the latest data issued by the Treasury, as of February 21, 2012, $10.76 billion[6] was awarded through the 1603 Treasury grant program to roughly 22,747 projects[7] (Department of Treasury 2012). As seen in the following figures, 22,060 approved solar projects received $1.99 billion (18%) of the total grant money awarded. Of this amount, $1.85 billion (93%) was awarded to "solar electric" projects and $133 million (7%) to "solar thermal" projects. Wind projects received the majority of Treasury grant awards (in dollar value) with over $8.15 billion, or 76% of the total. The remaining Treasury grants were awarded to an array of other renewable and advanced generating technologies.

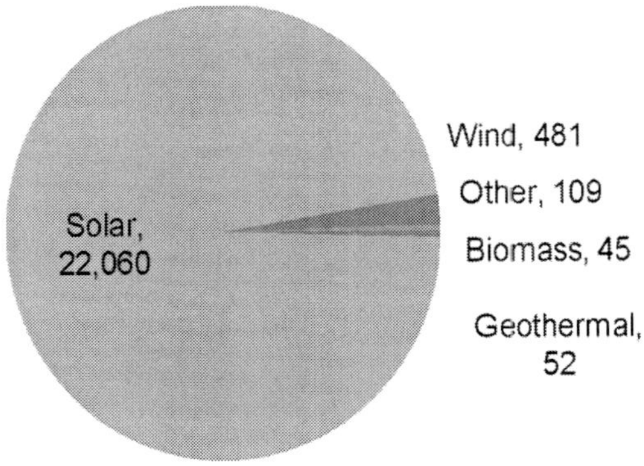

Source: Department of Treasury 2012.

Figure 1. Grants awarded by technology.

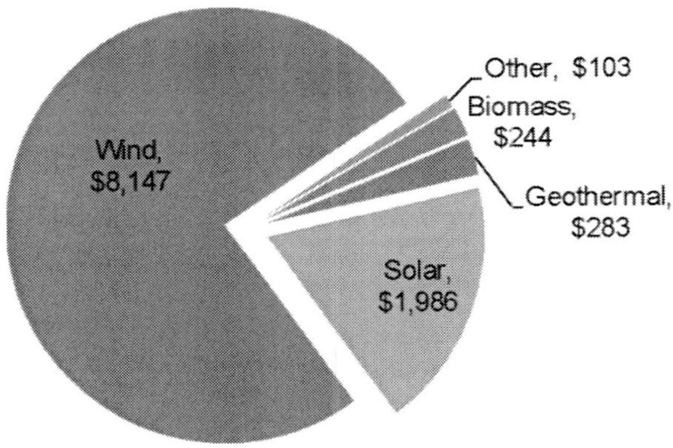

Source: Department of Treasury 2012.

Figure 2. Treasury funding by technology ($ millions).

Assuming the 1603 awards equal 30% of all capital invested,[8] the $10.76 billion awarded as of February 21, 2012, has induced $25.11 billion in private investment or led to a total investment of $35.87 billion in renewable energy projects since its inception in 2009.

The top 15 grant-receiving states are depicted in Figure 3 (Department of the Treasury 2012). Texas, with a substantial number of wind projects, leads the nation, receiving almost $1.7 billion in total Treasury grant dollars. Entities in all 50 states, the District of Columbia (D.C.), and Puerto Rico have been awarded Treasury grants.

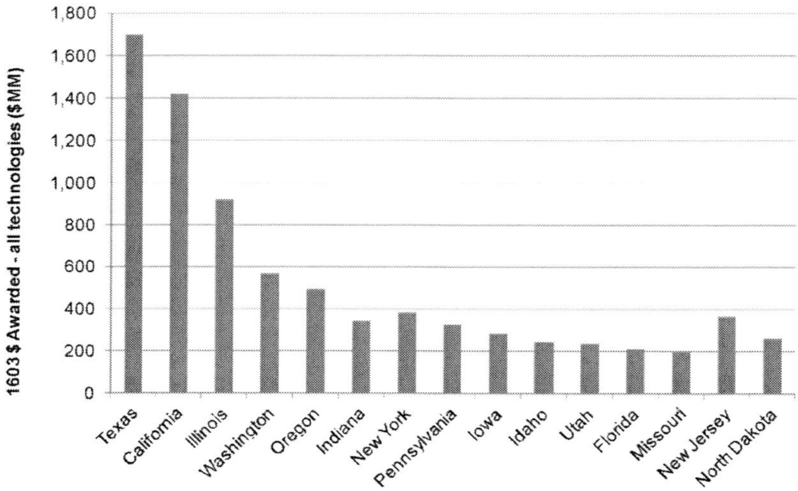

Figure 3. Treasury grants awarded for top 15 States (as of February 21, 2012).

Since its inception in 2009, the Treasury grant program has been widely credited with improving the speed and accessibility of financing for solar projects (Mendelsohn and Harper, forthcoming). NREL gathers information on project finance terms and gauges attitudes on the value of renewable energy policies through an effort called the Renewable Energy Finance Tracking Initiative (REFTI). A REFTI questionnaire conducted in March 2011 suggests that the renewable energy industry highly values the Treasury grant program. The majority of participants, 42 of 48 (89%), indicated the program is either "extremely" or "very" important to project development success (Figure 4). Participants in the questionnaire place a higher level of importance on Treasury grants than on state incentives or RPSs.

In a separate REFTI poll conducted by NREL in August 2010, 27 participants indicated that 64% of their projects completed in the prior 12 months were "dependent" on the Treasury grant, meaning they would likely not have been completed without the Treasury grant support.[9]

One reason for Treasury grant popularity is the reduced transaction cost to utilize this support structure versus the ITC. Generally, renewable energy projects and their developers do not have sufficient tax liability to take advantage of the ITC directly and therefore must seek a tax equity investor to monetize the benefit. Sheldon Kimber, Chief Operating Officer of Recurrent Energy, explained in an interview that monetizing the ITC via a tax equity investor can consume 15%–40% of the ITC benefit (Kimber 2010). Kimber explained that the high fees are not only due to the higher yield required by a tax equity investor to monetize a true tax credit but also because certain tax investors seek project ownership after the assumed 20-year life of the facility. Developers, on the other hand, assume cash flows to 25 or 30 years. Passing on project ownership—even after 20 years—alters the developer's return and, in essence, raises the transaction costs associated with monetizing tax credits.

Separately, in a third poll conducted by NREL, industry participants indicated that the transaction costs to monetize the ITC represented 23% of the benefit, roughly corroborating the value specified by Kimber (Mendelsohn 2010).[10] That is, on a weighted average basis, 23% of the benefit is lost to monetizing via a third-party tax equity investor. In contrast, poll participants indicated the transaction costs to monetize a Treasury grant equaled 14% of the benefit. Monetization of the Treasury grant can be completed in the form of bridge finance until the award is received.

If accurately represented by these polling participants, monetizing the ITC and the Treasury grant incurs a differential in transaction costs of almost 9%. This transaction cost differential is worth over $900 million of the $10.76 billion awarded via the 1603 program as of February 21, 2012. The estimated $900 million in savings represent a direct investment into renewable energy projects rather than transaction costs to monetize the federal benefit.

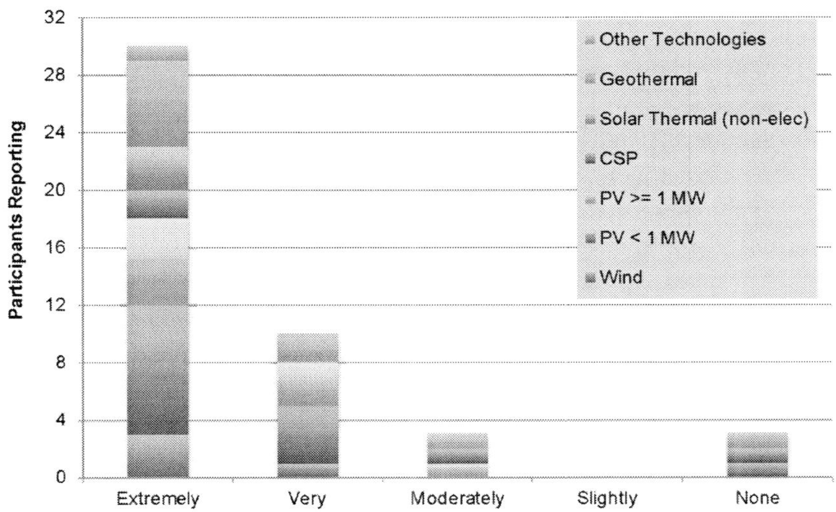

(Total participants = 48) Source: REFTI 2011.

Figure 4. Reported level of importance of the Treasury grant program by renewable energy technology.

2.1.3. Subsidized Financing

The Recovery Act allowed businesses to capture the full value of the ITC or Treasury grant even if projects receive subsidized energy financing (Bolinger et al. 2009). Prior to the Recovery Act, the ITC would not apply to the portion of the investment funded via subsidized financing, including below-market loans, state grants, and private activity bonds. This is frequently referred to as a reduced "basis," representing the percentage of capital expenditures that are eligible for recovery through the ITC. Now, the tax credit benefit of the project is not subject to a reduced basis. Accordingly, utility-scale solar facilities may be able to access tax-exempt funding sources such as municipal or state bonds or other low-interest loan programs without incurring a reduction in the ITC received.

2.1.4. Clean Energy Loan Guarantee Program

The Department of Energy (DOE)[11] offered three financial support structures administered by DOE's Loan Program Office. The loan programs were designed to "mitigate the financing risks associated with clean energy projects, and thereby encourage their development on a broader and much-needed scale" (DOE 2011). These were:

- Section 1703: Section 1703 of Title XVII (from the Energy Policy Act of 2005) authorizes DOE to guarantee loans for projects that employ new or significantly improved energy technologies and avoid, reduce, or sequester air pollutants or greenhouse gases.
- Section 1705: Section 1705 of Title XVII, added by the Recovery Act, authorizes DOE to guarantee loans for certain clean energy projects that commenced construction on or before September 30, 2011. The program was suspended after that date.
- Advanced Technology Vehicles Manufacturing: Section 136 of the Energy Independence and Security Act of 2007 authorizes DOE to provide direct loans to finance advanced vehicle technologies.

Only 1703 and 1705 are relevant to utility-scale solar. Solar projects (both generation and manufacturing) have only been awarded loan guarantees under 1705.

Through the 1703/1705 programs, the government guarantees debt associated with new energy production or manufacturing facilities relevant to renewable and other energy technologies (e.g., clean coal). Government guarantee on debt lowers the risk and required yield on the funds raised for potential lenders (DOE 2011). Initially under the Recovery Act, Congress appropriated $6.0 billion to the 1705 program. However, in August 2009 $2.0 billion was redirected to the Car Allowance Rebate System (also known as Cash for Clunkers program), and in August 2010, an additional $1.5 billion was removed to pay for extended unemployment benefits and other initiatives. These withdrawals have reduced the total 1705 program size to approximately $2.4 billion (Bartlett and Goldstein 2010).

Loan guarantees are awarded in the form of credit subsidies (Jaffe 2009). A credit subsidy is a similar structure to a loan loss reserve, in which banks set aside money to cover defaults in their loan portfolio. Likewise, for each loan guarantee award, the federal government sets aside a sum (the credit subsidy) in the project's name, which acts as insurance in the case of project failure. The credit subsidy is specific to each loan but was initially estimated by industry experts to average roughly 10% (Bracewell & Giuliani 2009). That means every dollar spent in credit subsidy would leverage $10 in loans.

The total investment supported by the credit subsidy allocation, including investor equity, could actually be much larger than indicated. For example, assume an investment was financed with 60% debt and 40% equity. In such a scenario, assuming an 11.7% credit subsidy, every dollar of credit subsidy invested could support $8.55 in debt and $5.70 of equity, or $14.25 in total investment (see Figure 5). In this case, the 11.7% credit subsidy represents only 7% of the total capital invested. The equity investments of the projects under the 1705 program are unknown, but this concept illustrates the potential capability of the credit subsidy to induce capital investments.

Table 1. 1705 Loan Guarantee Information

Program	Loan Amount ($billions)	Estimated Credit Subsidy at 11.7% Rate ($billions)
PV Electric Generation	6.14	0.72
CSP Electric Generation	5.86	0.69
Solar Manufacturing	1.28	0.15
Wind Generation	1.70	0.20
Other Technologies	1.19	0.14
Total	**16.16**	**1.88**
Solar Electric Generation Only	*12.00*	*1.40*

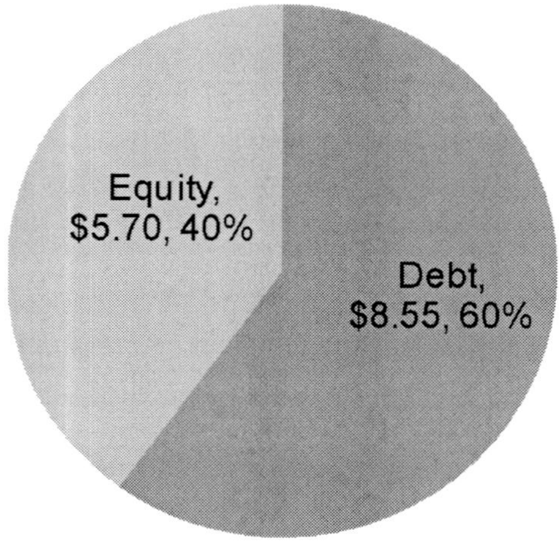

Figure 5. Hypothetical financial structure from $1 of loan guarantee.

The solicitation for loan guarantees for "innovative" technologies (Section 1703) was issued on July 29, 2009. "Innovative" technologies are defined as those that either (1) have not been installed and used in three or more projects in the United States or (2) involve meaningful and important improvements in productivity or value in comparison to commercial technologies (Klepper 2009). Loans are available from the Federal Financing Bank—which can provide very low-cost debt—if an applicant seeks a 100% guarantee from DOE (WSGR 2009).

The solicitation for loan guarantees for "commercial" technologies (Section 1705) was issued on October 7, 2009 (DOE 2010). The solicitation defines "commercial" technologies as those that have been used in three or more commercial applications anywhere in the world and that have been in operation for at least two years (Klepper 2009). Under this solicitation, DOE required that developers also partner with a private lender. DOE created the Financial Institution Partnership Program (FIPP), which specifies debt cannot exceed 80% of total capital. The DOE guaranteed up to 80% of the loan, or 64% of the total capital (DOE 2010). FIPP was designed for simple project finance structures without complex tax equity arrangements (Klepper 2009). Accordingly, these projects—which may not have completed financing as of this report publication—will likely not be able to use a complex financial structure such as a partnership flip or sale-leaseback to monetize the tax benefits and take advantage of the loan guarantee program. These financial structures are discussed in more detail in Section 3.2.

Table 2. Renewable Energy Loan Guarantee Awarded (by Program)

Program	Loan Amount ($billions)	MWAC	Exp. Annual Generation (GWh)
PVElectric Generation	6.14	2,272	4,730
CSP Electric Generation	5.86	1,243	3,623
Solar Manufacturing	1.28		
Wind	1.69	1,025	2,188
Other Technologies	1.19	180	1,492
Total	16.16		
Solar Generation Only	**12.00**	**3,515**	**8,353**

Table 3. Solar Electric Generation DOE Loan Guarantees Awarded (by Project)

Technology/Project	Loan Amount ($B)	MWAC	Exp. Annual Generation (GWh)
PVElectric Generation			
First Solar, Inc. (Antelope)	0.65	230	623
First Solar, Inc. (Desert Sunlight)	1.46	550	1,280
Prologis	1.40	752	1,015
SunPower (CVSR)	1.24	250	740
Mesquite Solar 1, LLC (Sempra Mesquite)	0.34	170	350
Cogentrix	0.09	30	75
Agua Caliente	*0.97*	*290*	*648*
Sub-Total PV Elec. Generation	6.14	2,272	4,730
CSP Electric Generation			
SolarReserve, LLC (Crescent Dunes)	0.74	110	504
Abengoa Solar, Inc. (Mojave Solar)	1.20	250	617
NextEra Energy Resources, LLC (Genesis Solar)	0.85	250	560
BrightSource Energy, Inc. (Ivanpah Solar Gen. System)	1.63	383	998
Abengoa Solar, Inc.	1.45	250	944
Sub-Total CSP Elec. Generation	5.86	1,243	3,623

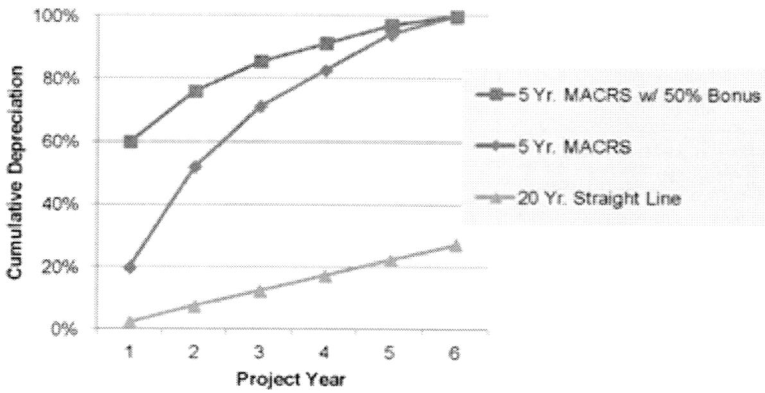

Figure 6. MACRS, bonus MACRS, and 20-year straight-line depreciation.

To date, renewable energy projects have only been awarded loan guarantees under the 1705 program. In total, the 1705 program awarded $16.16 billion in loan guarantees to 28 projects (DOE 2011). Of those, 19 are power generation projects, including PV, CSP, wind, and geothermal. Solar generation represents 12 projects (7 PV, 5 CSP) expected to produce a peak output of over 3,500 MW, supported by $12.0 billion in loan guarantees.[12]

The specific solar generation projects awarded 1705 loan guarantees are shown in Table 3.

Continued access to the loan guarantee program is a concern to developers of large projects. According to David Crane, President and CEO of NRG Energy, Inc., projects similar to the 290 MW Agua Caliente project or the 250 MW Colorado Valley Solar Ranch—both supported by the loan guarantee program—would not be viable in the current financial environment without loan guarantees (Engblom 2011).

2.1.5. Modified Accelerated Cost Recovery System Depreciation

Another federal policy encouraging utility-scale solar is Modified Accelerated Cost Recovery System (MACRS). Under Section 168 of the tax code, "equipment which uses solar energy to generate electricity" qualifies for five-year, double declining-balance depreciation (GE Capital 2010). As shown in Figure 6, MACRS allows full depreciation of the asset over five years of project life, or six tax years. Depreciation of the asset can be further accelerated through "bonus depreciation," which provides significant tax benefits in the first year of an asset life. The bonus depreciation reverts from 100% of the eligible plant in 2011 to 50% in 2012. These depreciation schedules are compared to a 20-year straight-line schedule (over the first six years only) to show the relative speed at which capital can be reduced in book value in order to lower taxes payable on income.

The five-year MACRS depreciation schedule provides a tax benefit equal to about 26% of system costs on a present value basis (Bolinger 2009a). In comparison, straight-line depreciation provides a tax benefit value of 14% of system costs. When combined, the accelerated depreciation and the 30% ITC provide a combined tax benefit equal to about 50%–60% of the installed cost of a commercial PV system (Mendelsohn et al. 2012).

2.2. Renewable Portfolio Standard Policies

State-level RPS policies[13] are a significant driver for utility-scale solar development, especially in areas with good solar resources. An RPS requires electric utilities or load serving entities (LSEs) to source a percentage of their electric load from renewable electricity generation. These targets are typically expressed as a percentage of total electricity consumption and range from approximately 2% in Iowa[14] to 33% in California (DSIRE 2012a). As of January 2012, 29 states, D.C., and Puerto Rico have established mandatory RPS policies and eight more have established non-binding renewable portfolio goals (RPGs).

Mandatory state RPS policies have been particularly successful at advancing new renewable energy development in a number of states (Wiser and Barbose 2008). Renewable energy goals are non-binding targets either established by legislation or executive order (Cory and Swezey 2007).

2.2.1. Solar Provisions in State RPS Policies

An increasing number of states have adopted distributed generation (DG), solar set-aside, or credit multiplier provisions in their RPS policies to provide differential support to promising technologies that are currently of higher cost (Wiser and Barbose 2008). A solar set-aside stipulates that a portion of the annual renewable energy compliance requirements be fulfilled with solar electricity. Credit multipliers give favored technologies more credit towards meeting RPS requirements than other technologies.

As of January 2012, solar provisions (excluding limited eligibility for DG-specific applications) have been implemented in 16 states and D.C. (DSIRE 2012b). Figure 8 shows the specifics of the solar and DG provisions currently implemented.

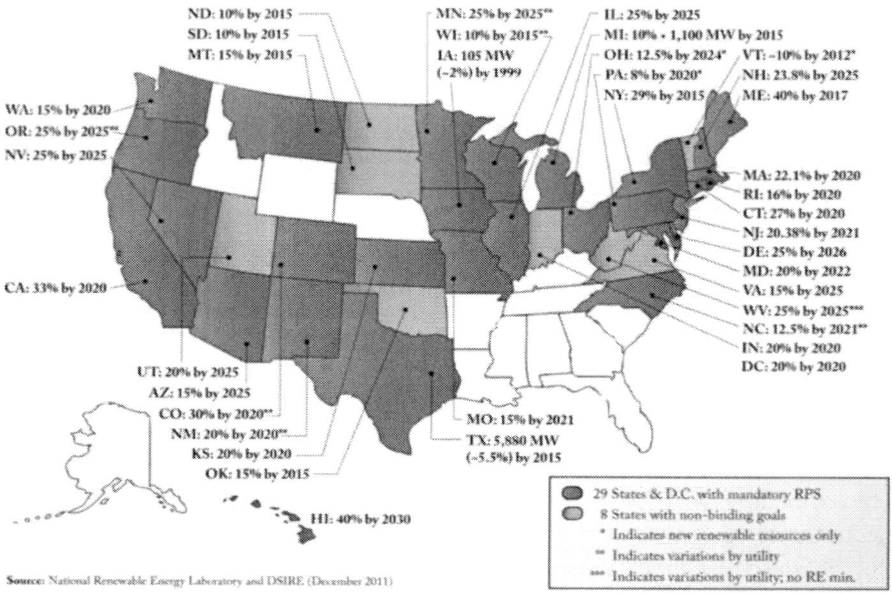

Figure 7. States with RPSs and non-binding goals.

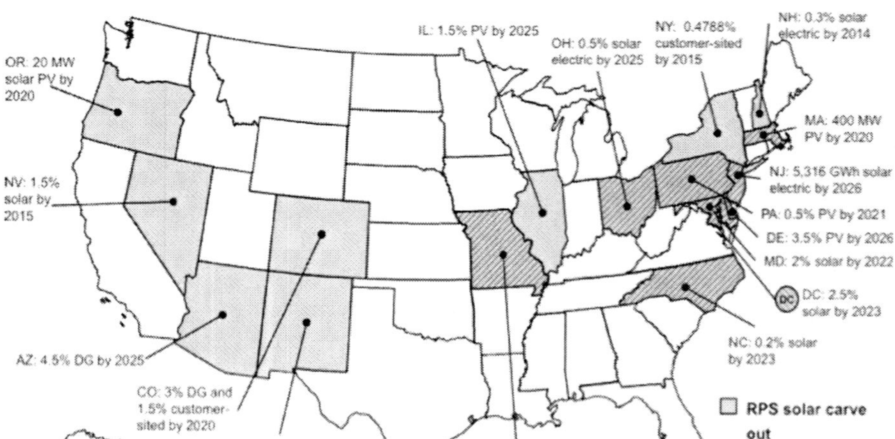

Source: Heeter and Bird 2011.

Figure 8. Solar and DG provisions in state RPS policies.

Table 4. Multipliers in State RPSs and RPGs (Excluding DG Multipliers)

Solar PV	Solar Electric	Various Multipliers (for which solar is eligible)	Non-Wind (includes solar)
Oregon (2x)	Utah (2.4x) (RPG)	West Virginia (RPG) (3x for renewable energy on a reclaimed mine)	Texas (2x)
Nevada (2.4–2.45x)	Michigan (3x)		
Delaware (3x)	Colorado (3x) (for publicly owned utilities only)	Colorado (1.5x) (community-owned renewable energy, < 30 MW, publicly owned utilities only)	

Source: DSIRE 2012b; DSIRE 2011b.

Credit multipliers are an alternative approach used by state policymakers to incentivize investment in one or more renewable technologies. Credit multipliers for solar have ranged from 1.1 to 3.0 times the credit that other renewable energy technologies receive (DSIRE 2012b). As shown in Table 4, six states have specific solar energy multipliers. Additionally, Texas has a multiplier for non-wind generation, and West Virginia has multipliers for renewable energy (including various alternative energy sources). Projects developed in the service territories of publicly owned utilities in Colorado receive a 3x credit multiplier for solar generation until July 1, 2015, or a 1.5x credit multiplier for community-owned renewable energy generation (DSIRE 2011b).

2.2.2. State and Regional Renewable Energy Certificate (REC) Markets

RECs, also known as green certificates or green tags, represent a quantifiable mechanism by which to reward renewable energy projects through an additional revenue stream. These certificates represent the environmental, social, and other non-power attributes of renewable electricity generation (Holt and Bird 2005). To comply with RPS policies, utilities or LSEs can directly own renewable energy generation, purchase renewable electricity bundled with RECs through PPAs, and/or purchase unbundled RECs separately (e.g., procured via separate contracts) (Holt and Wiser 2007).

When a utility negotiates a bundled contract with a developer, the parties agree upon the contract duration and PPA price. Under a typical PPA, the buyer agrees to purchase some (or all) of the project's generation at a fixed price or at a price with an escalation term (Cory et al. 2008). The PPA price usually includes an energy payment (dollars per megawatt-hour of generation) and a REC payment (dollars per megawatt-hour). Some PPAs will also include a monthly or annual capacity payment (dollars per installed kilowatt), but this is less frequent for variable generation technologies, such as solar, that lack storage capability.

To show compliance, utilities or LSEs can, in many states, purchase unbundled RECs through long- or short-term contracts or on the spot market (usually to true-up the amount of renewable electricity procured in a given year). Long-term REC contracts provide the most revenue certainty to investors in renewable energy projects.

Because of their value to utilities to demonstrate compliance, RECs can be an important source of revenue for project developers. However, project developers may find that risk-averse investors require long-term contracts with a utility for RECs at predefined prices. Without long-term contracts or other mechanisms of risk mitigation, investors may discount the value of a developer-projected future solar REC (SREC) stream by 50%– 90% (Cooney et al. 2008).

Prior to the 2008 financial crisis, some equity investors were willing to finance the construction of renewable energy plants that did not have long-term contractual agreements like PPAs (Schwabe et al. 2009). This is sometimes referred to as the "merchant power" model, whereby generators sell RECs and/or electricity to power marketers on the spot market or on a short-term contract basis, which permits them to take advantage of any upside.

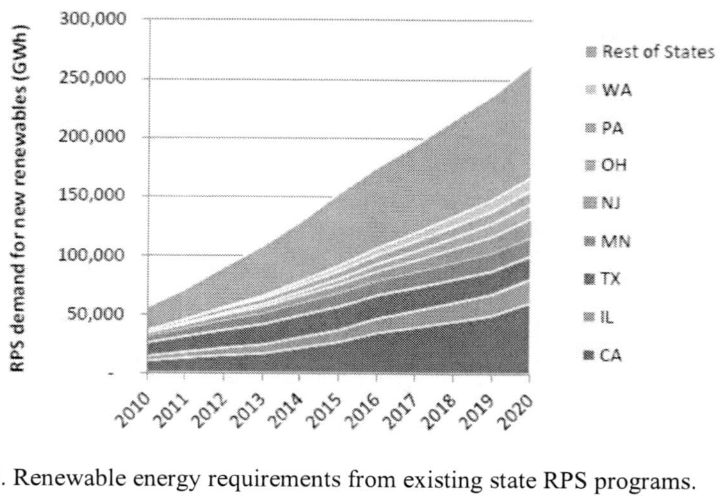

Figure 9. Renewable energy requirements from existing state RPS programs.

2.2.3. Solar Capacity Required to Meet RPS Requirements

While state RPS and solar set-asides have been instrumental in wide-scale procurement and deployment of solar power, recent evidence indicates the solar industry can already develop solar power projects at rates necessary to meet future RPS requirements. Accordingly, current state RPS regulations are expected to be less of a stimulus to solar power expansion in the future.

The Lawrence Berkeley National Laboratory recently estimated that nine states and D.C. have specific SREC requirements and will require incremental solar capacity of 6,729 MW by 2025 (Barbose 2010).[15] That represents roughly 480 MW of solar capacity installed per year, less than half the total domestic installations in 2010.

Of the nine states and D.C. that have SREC requirements, New Jersey is the most aggressive. The state will require over 3,700 MW of incremental solar capacity by 2025 over its current requirement of 319 MW, equal to 55% of all states with SREC requirements combined (Barbose 2010). While New Jersey's RPS dictates more than a 10-fold build-out over its current solar capacity, the value represents only 264 MW per year, or 66 MW per quarter. To put that in perspective, the state achieved 64 MW of solar installations in Q3 2011, thus currently achieving the pace required through 2025.

Overall, state RPS requirements (for all eligible renewable technologies) will necessitate production of approximately 260,000 GWh in 2020, roughly quadruple the 55,000 GWh of renewable energy required in 2010. On average, the RPS will require an additional 20,000 GWh produced each year through 2020.

To comprehend the overall RPS requirement of an additional 20,000 GWh per year, the hypothetical analysis in Table 5 was developed. In this hypothetical, all solar resources (PV and CSP) were assumed to provide 20% of the total RPS energy and generate at an average capacity factor of 20% as a crude approximation. The analysis indicates 2,283 MW of solar resources would be required. That value represents 570 MW per quarter, or about 27% more than the quarterly PV capacity installed in Q3 2011.

2.2.3.1. RECs in Wholesale Power Markets

RECs are more frequently unbundled from electricity in regions where independent system operator (ISO) wholesale power markets have developed (generally in states where electricity market restructuring occurred). REC trading, facilitated by brokers, occurs most frequently in Texas, the Mid-Atlantic (PJM Interconnection), and New England (ISO-NE) but is permitted in numerous state RPS policies (Holt and Bird 2005).

Table 5. Potential Renewable Energy Annual Capacity Additions

Technology	% of Assumed Portfolio	Annual Energy Produced	Assumed Capacity Factor	Avg. Annual Capacity Addition (MW)
Wind	70%	14,000	35%	4,566
Solar	20%	4,000	20%	2,283
Biomass	5%	1,000	80%	143
Geothermal	*5%*	*1,000*	*90%*	*127*
Total	100%	20,000		7,119

Markets specifically for SRECs are developing in states where policymakers have established solar set-asides within their RPS requirements and have set high solar alternative compliance payments (ACPs). An ACP is a penalty payment for failure to meet an RPS requirement in a given year. In fact, the most active REC spot markets are those where RPS penalty provisions are priced higher than the actual cost to develop eligible projects (Cory et al. 2008a).[16]

Since 2009, SREC prices dropped considerably in most markets (Heeter and Bird 2011; see Figure 10). The New Jersey SREC market, the largest in the country, has experienced spot market SREC price drops of roughly two-thirds in the six months ending September 2011, representing a trend that supply is outpacing demand (Heeter and Bird 2011).

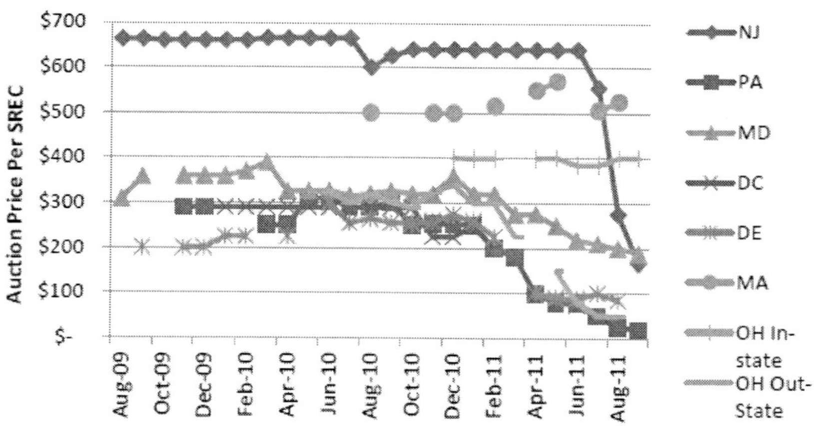

Figure 10. SREC spot prices (per MWh), August 2009–September 2011.

2.2.4. Focus on the U.S. Southwest Solar Market

The U.S. southwest—the territory encompassed by Arizona, California, Colorado, Nevada, New Mexico, and Utah—has the strongest solar resource of any region in the country (both for CSP and PV). With aggressive renewable energy targets, and perhaps the provision of solar set-asides, the states in this region could potentially drive the growth of the U.S. solar market.

Five of the six southwestern states have enacted an RPS, and the remaining state, Utah, has an RPG, a non-binding requirement that utilities must fulfill to the extent it is cost effective (DSIRE 2012a).The targets for investor-owned utilities (IOUs) and relevant solar provisions are bulleted in Figure 11.

New Mexico and Nevada are currently the only states in the Southwest with solar provisions in their RPS. If these provisions are met, then these two states could be supplied with almost 1,500 GWh of solar generation annually by 2025 (Barbose 2010).[17]

Even in the absence of a specific solar set-aside, market demand for solar electricity is expected to increase in other southwestern states due to the quality of the solar resource and the lack of other renewable resources. For example, the "Arizona Renewable Energy Assessment" conducted by Black & Veatch estimates that in 2025, 65% of the Arizona renewable demand (or over 8,000 GWh) will be met by solar (Black & Veatch 2007). Assuming a 20% capacity factor, roughly 4,500 MW of solar resources could be installed in Arizona to meet in-state RPS requirements.

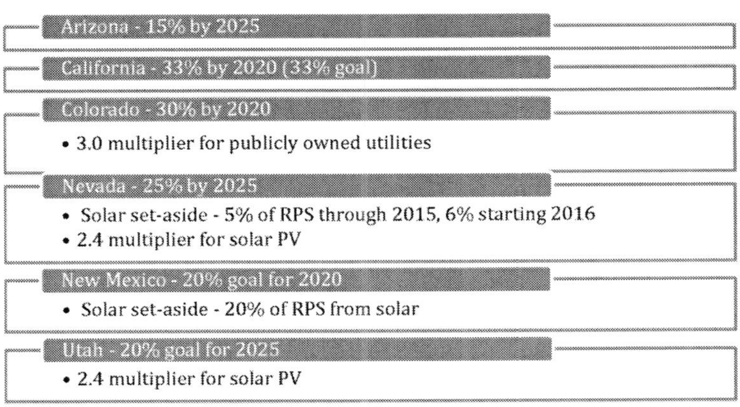

Source: DSIRE 2012a.

Figure 11. RPS provisions in southwestern states (IOUs only).

Table 6. Projected Oversupply or Unmet Demand for Renewable Energy Capacity in 2020

				Arizona	California	Colorado	Idaho	Montana	Nevada	New Mexico	Oregon	Utah	Washington	Wyoming	**Total WECC**
SUPPLY	As of December 2011 – Not Including Hydropower[4]		Existing Renewable Energy Plant Capacity (MW)[1] (SNL 2011c)	194	7,880	1,372	441	404	589	738	2,558	287	2,547	1,419	**18,429**
			Planned Renewable Energy Projects Capacity (MW) (SNL 2011a)	1,045	10,367	91	564	300	1,374	130	864	0	1,245	2,841	**18,820**
			Sum of Existing and Planned Renewable Energy Capacity (MW)	1,239	18,247	1,462	1,005	704	1,963	868	3,422	287	3,792	4,260	**37,249**
DEMAND	Capacity Required in 2020 to Meet Current RPS Requirements (MW)[2]	**A**	High Capacity Factor, Low Capacity (MW)	1,380	18,653	2,741	0	325	958	896	1,733	0	1,807	0	**28,493**
		B	LBNL Capacity Factors (Barbose 2011) (MW)	1,415	27,944	2,741	0	382	896	985	1,811	0	2,952	0	**39,126**
		C	Low Capacity Factor, High Capacity (MW)	2,144	31,354	2,741	0	662	1,421	896	3,079	0	3,238	0	**45,535**
GAP	Oversupply or (Unmet Demand) in 2020[3]	**A**	High Capacity Factor, Low Capacity (MW)	(141)	(406)	(1,279)	1,005	379	1,005	(28)	1,689	287	1,985	4,260	**8,756**
		B	LBNL Capacity Factors (Barbose 2011) (MW)	(176)	(9,697)	(1,279)	1,005	322	1,067	(117)	1,611	287	840	4,260	**(1,877)**
		C	Low Capacity Factor, High Capacity (MW)	(905)	(13,107)	(1,279)	1,005	42	542	(28)	343	287	554	4,260	**(8,286)**

Source: Haase et al. 2012.

A recent analysis of the 11 states that comprise the western interconnection system, however, indicates only 1,877 MW of renewable capacity are needed in the broader region by 2020 beyond the slate of currently planned facilities (Haase et al. 2012)[18] (see Table 6). Importantly, the analysis assumes significant transmission capacity will be built to interconnect new renewable energy facilities in Wyoming, Oregon, Idaho, and Washington with California, Arizona, and Colorado, which are projected to have unmet demand for renewable energy.

2.3. State-Level Tax Incentives

Utility-scale solar systems may be eligible for multiple types of state tax incentives. Developers often use state tax incentives in concert with other incentives to support renewable energy deployment (Lantz and Doris 2009). The primary state tax policies that apply to large-scale solar energy development are corporate, property, and sales tax incentives. Corporate income tax incentives may take the form of tax deductions or tax credits and may be awarded on the basis of investment, or they can be based on the quantity of renewable electricity produced.

State property tax incentives reduce or limit the amount of taxes levied against property improvements as a result of building renewable energy projects. Sales tax incentives reduce—or provide exemptions from—sales and use taxes levied against goods purchased for renewable energy projects (Lantz and Doris 2009).

As of December 2011, in various forms, 16 states offer corporate tax incentives, 31 states offer property tax incentives, and 21 states offer sales tax incentives for solar projects (DSIRE 2012c; DSIRE 2012d; DSIRE 2010e).

2.4. State-Level Clean Energy Funds

Clean energy funds are typically developed to ensure continued support for renewable energy, energy efficiency, or low-income energy programs. Two common uses are to support rebate programs for renewable energy systems and to fund renewable project loan programs. Clean energy funds are complementary to other renewable energy support policies on the state and federal levels. Any financial support received through a state fund may increase the effectiveness of other policies (e.g., tax incentives or RPS) by bringing down capital costs.

As of January 2012, 18 states, D.C., and Puerto Rico have implemented renewable energy technology funds, generally referred to as public benefit funds or system benefit funds programs (DSIRE 2012f). The funds are most often supported by a small surcharge on electricity consumption but can also be based on voluntary funding (as in Maine) or recharged through loan repayments and other returns on investment (as in Pennsylvania) (Doris et al. 2009).

According to Clean Energy Group, between 1998 and 2007, state clean energy funds invested approximately $1.5 billion in renewable energy projects and leveraged an additional $2.6 billion in private capital (CESA 2009). State funds have supported a broad range of renewable technologies including significant investment in grid-connected solar PV. States traditionally rely on RPS policies to promote large-scale development and use the clean energy funds to encourage DG projects. It is estimated that between 1998 and 2007, 82% of the financial support from state clean energy funds was allocated to distributed energy projects, whereas 18% was allocated to grid-scale projects (CESA 2009). Although likely limited in their application to utility-scale solar systems, clean energy funds should be investigated as a potential source of project capital.

In addition to providing direct financial support to renewable energy, clean energy funds may be designed to fund initiatives with longer-term benefits, such as research, development, and demonstration, or initiatives that address regulatory or market barriers. Furthermore, clean energy fund investments can stimulate the development of companion industries and infrastructure and promote consumers' awareness of clean energy (Doris et al. 2009).

3. Financing Mechanisms to Develop Utility-Scale Solar

3.1. Background

Many of the policies supporting renewable energy comprise tax incentives (in the form of credits or accelerated depreciation benefits), as discussed in Section 2. However, renewable energy projects and their developers generally do not have enough tax liability, also known as "tax appetite," to utilize the credits to their full potential. The tax credits can be "carried forward" to shelter future income, but this greatly reduces the present value of the benefit.[19]

Accordingly, a number of financial structures have been developed to take advantage of the tax credits at their highest value. These structures generally require an equity investment from a firm with sufficient tax appetite to utilize the tax credits; this is referred to as a tax equity investment. The firm is therefore the tax equity investor. Such investments, and the financial structures that build upon them, are designed to maximize the value of federal, state, local, and utility-based incentives; allocate risk and reward among different funding sources; and allow project participants to focus on core competencies.

According to a database of questionnaire responses via NREL's REFTI project,[20] financing-related aspects of project development were referenced as the largest barrier to project development for 52% of respondents in large-scale PV projects (>1 MW) and 41% of respondents involved in CSP projects.[21] Figure 12 provides the questionnaire responses for the nearly two years of data collected from Q4 2009–1H 2011.

Notably, REFTI participants involved in PV and CSP deployment offered distinct viewpoints on the primary barriers to successful project development. About 21% of those involved in PV projects indicated that finding tax equity was a critical barrier; by contrast, 0% of CSP respondents considered tax equity a barrier. About 20% of this group indicated instead that access to government incentives was their primary concern.

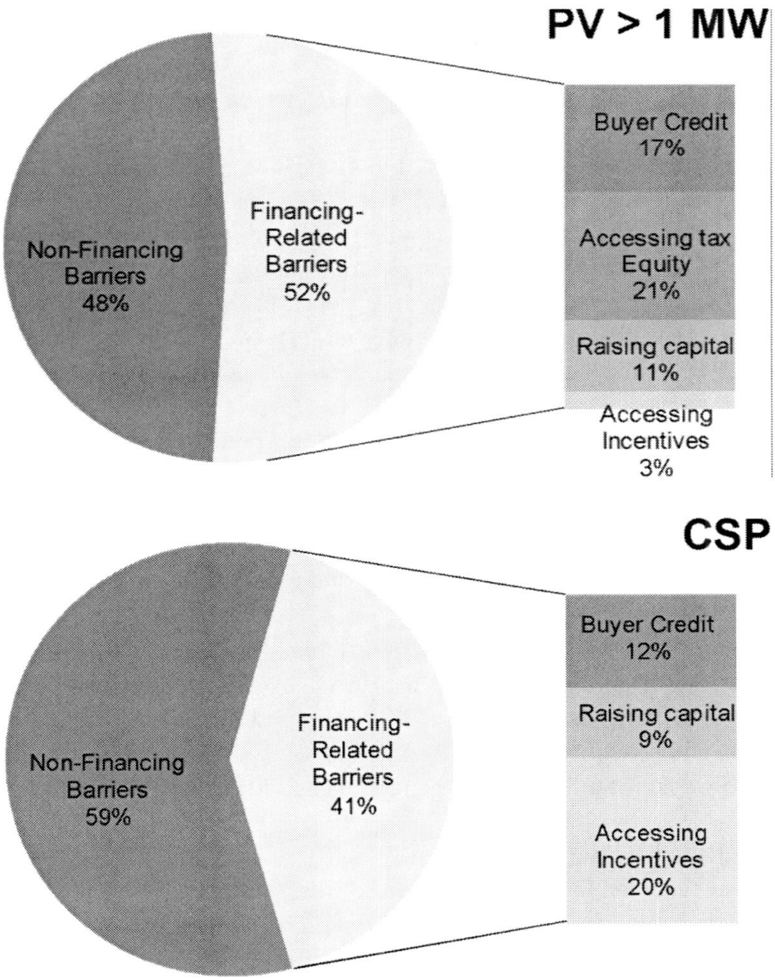

PV (n = 115) and CSP (n = 34).

Figure 12. Primary barriers to renewable energy project development.

Of the non-financing barriers referenced by REFTI participants, project economics and transmission had the highest reporting. PV respondents, in particular, indicated access to transmission as their primary barrier. Participants involved in CSP projects also listed technological factors as critical to project success.

3.2. Specific Financing Structures

A variety of project finance structures may be applied to utility-scale solar. This report discusses four structures that may be employed in the market: (1) single owner (also referred to as balance sheet finance); (2) leases, including sale-leaseback and inverted lease; (3) partnership flip; and (4) utility prepay.[22] An array of adjustments can be made under each structure based on the needs of a given project. Moreover, projects may employ unique, one-off financing structures based on the risk appetites of the entities involved (including the developer, investor, and utility) as well as the investment climate at the time (e.g., the cost of debt and equity and perception of technology risk).

These structures are born out of the current ITC and accelerated depreciation structures. If the ITC expires at the end of 2016, as currently legislated, these structures may lose some appeal. The DOE and NREL are currently evaluating how projects might be financed without the ITC and gaining valuable examples in the securities and real estate industries.

3.2.1. Single Owner (Balance Sheet) Finance

Single owner finance represents the direct investment by a developer into a project asset using the balance sheet (as opposed to project finance) of the developing entity. Utility- scale solar projects can be balance-sheet financed in one of two ways: (1) utility-owned generation and (2) developer balance sheet.

3.2.1.1. Utility-Owned Generation

Owning generation projects allows utilities to diversify their portfolios and maintain a higher level of control over project siting and grid reliability impacts (especially for distributed projects). Utilities may experience shortfalls in renewable energy procurement to meet RPS targets if independent power producers (IPPs) cannot fulfill contractual obligations. Furthermore, as discussed in Section 2.1.1, with the passage of the EESA in 2008, investment in solar generation projects has become more attractive to utilities due to their ability to access the 30% ITC.

In utility-owned generation, utilities either purchase turnkey projects from independent developers pursuant to a competitive solicitation or they develop and build the project themselves (see Figure 13).

Though utilities have not traditionally invested in solar facilities, several factors make them well-positioned to invest in large electricity generation projects. IOUs are considered stable, creditworthy entities and can attract capital at a favorable interest rate (even in tight credit markets). IOUs also have franchised authority to provide electric services and are experienced in raising capital in both the public equity and debt markets. Accordingly, utility balance sheets could prove a valuable anchor to support solar project development.

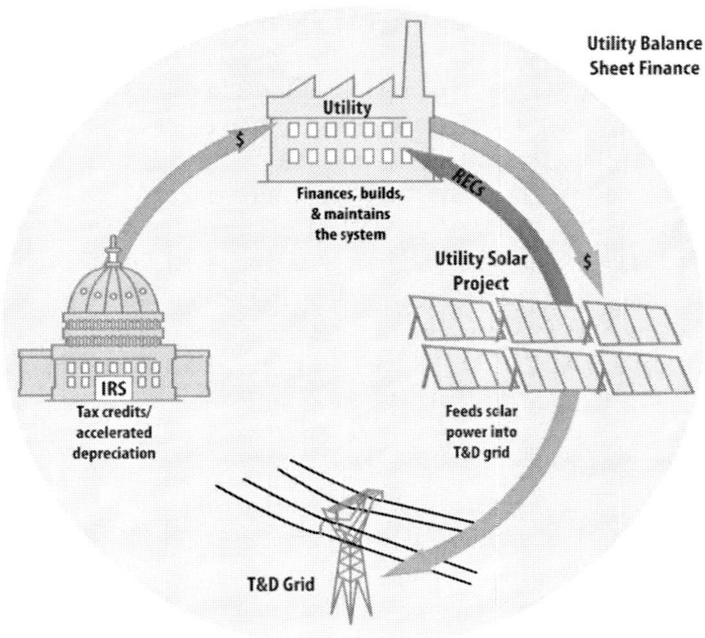

Figure 13. Utility balance sheet finance.

Utilities can invest in solar assets in one of four ways. First, utilities can directly finance, own, and operate solar energy projects for a variety of operational and economic reasons. For example, Pacific Gas and Electric

(PG&E) and Southern California Edison invest directly in solar energy projects via the Solar Power Initiative approved by the California Public Utilities Commission. Through the program, each utility will own 250 MW of solar to be developed over a five-year period. The investments will be recovered through each utility's base rates and be allowed to earn the company's weighted average cost of capital (CPUC 2009).

The largest IOU serving New Jersey, Public Service Electric & Gas Company (PSE&G), has also started to develop and operate its own solar systems. In July 2009, PSE&G received approval for its $515 million "Solar 4 All" program where it will install, own, and operate 80 MW of solar by 2013 (PSE&G 2009a; PSE&G 2009b). PSE&G's program will double the amount of solar in New Jersey—a state that already ranks second in the country for installed solar power—highlighting the ability of utility-owned generation to drive solar market growth.

Second, utilities can invest through developer subsidiaries or affiliates. Several major utilities—including Duke Energy, San Diego Gas and Electric (SDG&E), and NextEra (formerly Florida Power & Light)—have development arms that are very active in the market and have installed solar and other renewable technologies across the country. MidAmerican Energy Holdings, owned by Warren Buffett's Berkshire Hathaway, is the corporate owner to MidAmerican Energy Company, a large utility serving Iowa and surrounding states, and MidAmerican Renewables, LLC, the unregulated owner of wind, solar, geothermal, and hydro power plants (MarketWatch 2012). MidAmerican made significant headlines in December 2011 when it purchased the 550-MW, $2.4 billion Topaz Solar Farm project from First Solar and a 49% share of the 290 MW, $1.8 billion Agua Caliente project from NRG (MoneyNews 2012). To finance the Topaz purchase, MidAmerican successfully offered $850 million of private placement bonds in February 2012. The 27.5-year, 5.75% bonds were rated BBB-, the lowest investment grade available, by Fitch Ratings. As of March 2012, a second bond offering is being planned to raise an additional $430 million.

Third, utilities can use ratepayer tax appetite to support investment by a third party. For example, in 2011, SDG&E invested tax equity into the Rim Rock Wind facility located in northern Montana (Schwabe 2011). The deal—approved by the Federal Energy Regulatory Commission in January 2012—invests $285 million of tax equity, most of which is ratepayer funded. SDG&E will purchase the energy and RECs, although the energy will not be delivered to the utility (Recharge 2012). The structure is a first for California utilities. Certain stakeholders to the regulatory submission originally objected to investment of only ratepayer funds, and ultimately negotiated with SDG&E to lower ratepayer investment in the project and add shareholder funds (Schwabe 2011).

Fourth, utilities can invest shareholder tax appetite into a third-party project. Such investment would not be recovered by ratepayers but could modify taxes paid by, and dividends paid to, shareholders. For example, subsidiaries of PG&E have invested tax equity in SolarCity and SunRun, enabling these entities to develop residential and commercial solar systems (PG&E 2010a; PG&E 2010b). The funds were provided purely as shareholder funds and will not be incorporated in rates.

However, the tax code treats utilities and third-party developers differently. "Normalization" accounting rules require utilities to levelize the ITC benefits when setting retail electricity rates. That is, the utility must spread the payment of the ITC benefit to ratepayers over the useful life of the asset.[23] Although the utility benefits from the ITC in the first year of the project, that benefit cannot be directly transferred to the ratepayers in the same year (Miller 2009). Because their use of the ITC cannot directly result in lower rates, electric utilities may incur a cost disadvantage relative to non-utility developers when all other things are equal.

3.2.1.2. Developer Balance Sheet

Developers often finance utility-scale solar projects by utilizing their own balance sheets. Some examples include:

- Duke Energy Generation Services (the unregulated developer subsidiary of Duke Energy) financed the Blue Wing Solar facility in Texas via balance sheet capital (Bloomberg New Energy Finance 2010a). The Blue Wing Solar project will consist of panels supplied by First Solar and is expected to generate about 14.4 MW (ac) at peak output. The power will be sold to CPS Energy under a 30- year PPA. Duke bought the project from juwi Solar in January 2010.
- NRG Energy used its balance sheet to finance the 13.5 MW second phase of the Blythe PV project. The project was originally developed by First Solar and later sold to New Jersey-based power generator NRG Energy in November 2009 (Bloomberg New Energy Finance 2010b).
- In October 2010, Iberdrola announced its intent to finance—via its balance sheet—construction of 50 MW of solar electric projects in Arizona and Colorado (Murray 2010). Construction costs were

reported to be $60 million. In November 2011, Iberdrola dedicated its 20 MW Copper Crossing Solar Ranch project located in Florence, Arizona (Iberdrola 2011). The project holds a 25-year power purchase contract with the local utility, Salt River Project.

A developer who maintains a strong financial position can recapitalize its balance sheet by selling an equity or debt position in their project or portfolio of projects (Kimber 2010). Once recapitalization takes place, the development cycle can start again. For example, Hudson Clean Energy financed Recurrent Energy's broader development capabilities by investing $75 million in 2008 (Recurrent Energy 2008). In return, Recurrent Energy completed the 5 MW Sunset Reservoir project in San Francisco and was able to sell project debt to recapitalize its balance sheet, thus enabling the development of other projects in its 1 GW portfolio pipeline.

3.2.2. Partnership Flip

Under a partnership flip structure, the project developer partners with a tax equity investor to fully use a project's tax benefits (Coughlin and Cory 2009). Partnership flip structures are specifically designed to take advantage of federal incentives, including the ITC and MACRS, which frequently cannot be fully utilized by the developer alone (Cory et al. 2009). Wind projects widely used this financial structure in the past.

The structure comes in two basic forms: the all-equity partnership flip, where all capital is provided as equity, and the leveraged partnership flip, where project-level debt is also provided by a lender. The debt in a leveraged partnership flip can be provided by the tax equity investor (as a separate tranche) or by a third entity depending on how the project is structured.

The flip structure is generally designed to provide the tax equity investor a pre-negotiated return in a set number of years (e.g., a 9% yield by year eight of the project). After that design goal is met, the annual stream of benefits (including tax benefits and cash) is reallocated, or "flips," to the sponsor to reward the risk taken and work invested. The partnership arrangement defines the terms between the developer and tax equity investor, including:

- Initial equity investments
- "Pre-flip" distribution of the project's cash revenues and tax benefits
- "Post-flip" distribution of the project's cash revenues and tax benefits.

The "flip-point" occurs once the tax equity investor achieves an agreed-upon internal rate of return associated with the equity contribution. Some important tax law considerations relevant to a partnership-flip structure include the following:

- The developer must have at least a 1% interest in all material items associated with the partnership
- Each tax equity investor must have at least a 5% interest in all material items associated with the partnership
- The tax equity investor must bear the risk of solar resource availability (another project participant may not guarantee the availability of the renewable resource) (Baker Botts 2007).

Figure 14 presents a potential tax equity partnership. In this example, the developer provides 40% of the initial equity with a tax investor providing the remainder. The benefits are divided into three streams: the cash (or net revenue from the project), the depreciation benefits from applying MACRS, and the ITC. In this example structure, the cash benefits are reallocated with two flip points, the depreciation benefits are reallocated with one flip, and the ITC benefits are not redistributed during the project life.

3.2.3. Lease Structures

Two basic lease structures are utilized by the U.S. solar industry. These structures include sale-leaseback and inverted lease.

3.2.3.1. Sale-Leaseback

In sale-leaseback financing, a project developer sells the project assets for cash and simultaneously signs a long-term lease with the investor. The financial transaction is accomplished by conveying the title of the asset, at an agreed upon value, to an investor in exchange for a lump-sum payment to the developer. The developer then makes lease payments to the investor in exchange for the cash injection (Jacobs 2009). A sale- leaseback is distinct from an inverted lease structure, also referred to as an operating lease, whereby the lessee has the option to purchase in year six or beyond (Kendall 2009).

The developer, in turn, arranges a PPA with the power purchaser, which becomes the primary revenue stream used to pay the lease payments. The developer (the lessee in this structure) is obligated to pay a fixed rent to the investor (the lessor) for the term of the lease regardless of how the system performs or if force majeure events occur (Feo and Tracy 2009). Figure 15 is a graphical representation of the sale-leaseback structure.

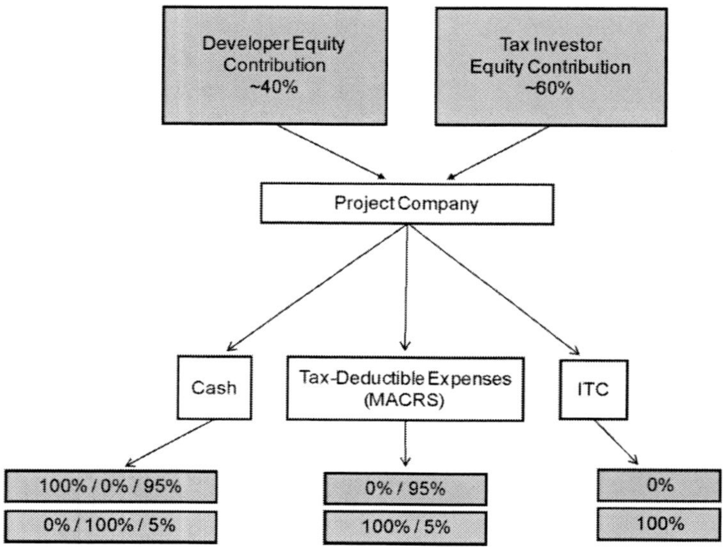

Source: Adapted from Harper et al. 2007.

Figure 14. Tax equity financial structure.

In June 2009, Wells Fargo and SunPower Corp. announced the development of a $100 million sale-leaseback financing program (SunPower 2009). The financing vehicle was later utilized to finance a 1 MW solar power system installed at the University of California, Merced (UC-Merced). Under the financing program used for the UC-Merced system, the developer (in this case, SunPower) entered into PPAs with qualified customers and designed, built, operates, and maintains the solar energy systems. Wells Fargo financed the project and receives the tax benefits. Customers hosting the systems buy the electricity from the developer with no initial capital investment (SunPower 2009).

A primary benefit of the sale-leaseback structure is that it provides 100% upfront financing. The sale-leaseback arrangement allows the business owner to raise capital while retaining the use of the assets that are needed in the business. Sale-leasebacks enable companies to realize the market value of their facilities and redeploy that capital into other projects (W.P. Carey 2010). A sale-leaseback can provide a significant source of funds for various purposes, including paying off a specific lender, buying back capital stock, buying out a partner, funding a working capital account, and/or upgrading assets (W.P. Carey 2010).

Sale-leaseback arrangements also allow a developer to strengthen its balance sheet by eliminating debt, increasing its borrowing capacity, and improving its return on assets or return on investment (Ray and Rawley 2007). Lease payments are considered operating expenses and thus are 100% tax-deductible. Sale-leaseback structures improve credit availability because capital equipment is no longer being financed at a regular bank, nor is it cutting into the lines of credit businesses retain with banks (Ray and Rawley 2007).

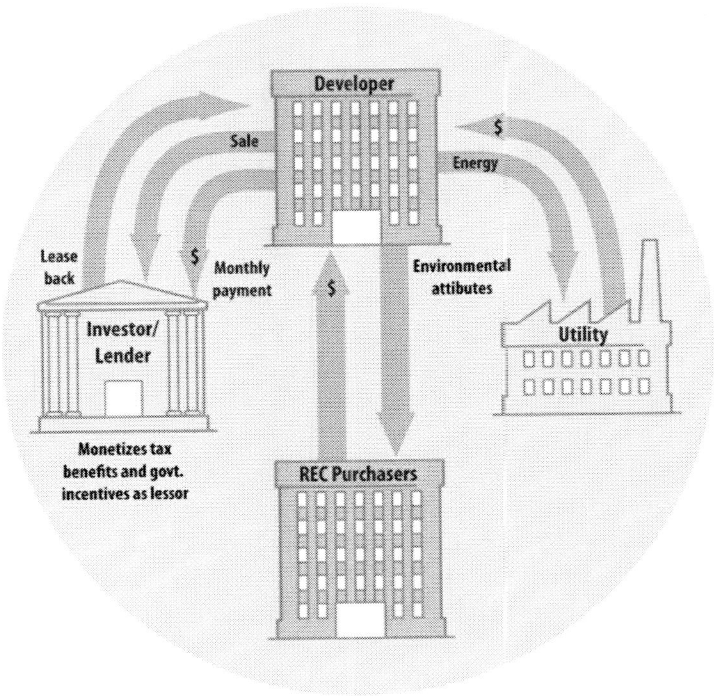

Figure 15. Sale-leaseback financial structure.

3.2.3.2. Inverted Lease

In an inverted lease (also referred to as a pass-through lease or master tenant lease), the developer leases the project to a tax equity investor and passes through the ITC to the tax equity investor. In turn, the tax equity investor, the lessee, sells the electricity to the developer via a PPA arrangement (Tracy et al. 2011). The developer may operate the system on behalf of the investor pursuant to an operation and maintenance agreement.

Unlike the sale-leaseback structure, the tax credit and depreciation benefits in the inverted lease are split between the lessor and the lessee (the developer retains them to reduce its taxes). The lease term in such transactions runs anywhere from 6 to 15 years (Competitive Energy Insight 2011). At the end of the lease, the developer takes back the project without having to pay any additional costs. According to some consultants, the market has utilized this structure "to an aggressive form that presents significant tax risk" (Competitive Energy Insight 2011).

The advantages of the inverted lease structure include an easy exit for the investor after the lease is completed, the ITC is based upon fair market value of the system rather than the developer's cost, and separation of the ITC from the depreciation benefits as some investors do not want both tax streams (Cuddy 2011).

A primary disadvantage of the inverted lease structure is that the investor/lessee is obligated to pay fixed rent to the developer/lessor for the term of the lease, irrespective of the actual performance of the system or existence of force majeure events, for example. The project must also be structured under specific guidelines for U.S. federal tax purposes. Finally, although not relevant to solar projects, the PTC cannot be combined with inverted lease structures because of tax-related ownership requirements.

3.2.4. Utility Prepay

Utility prepay (sometimes referred to as hybrid financing) uses a utility's low-cost capital to directly finance a renewable energy project. Under the structure, a utility will prepay for the energy to be delivered under a contract. In turn, the prepayment provides a massive down payment enabling the termination of the construction financing debt. The utility indirectly applies its access to low-cost capital to finance the facility and receives beneficially priced power in return (Delony 2008) (see Figure 16).

To date, only municipal utilities have taken advantage of this financial structure. This structure has also been limited to wind projects but remains a potential option for utility- scale solar.

In 2007, the Los Angeles Department of Water and Power (LADWP), along with the City of Burbank and City of Pasadena, signed a 20-year contract with UPC Wind for a 200 MW wind project being developed in Millard County, Utah (LADWP 2007). Acting on behalf of the LADWP and the municipalities, the Southern California Public Power Authority (SCPPA) prepaid for the energy to be delivered during the 20-year term with proceeds from an upcoming bond sale. By prepaying up to $270 million for a guaranteed amount of energy, the utilities saved at least $42 million compared to the cost of constructing and financing the plant themselves (Carnahan 2007).

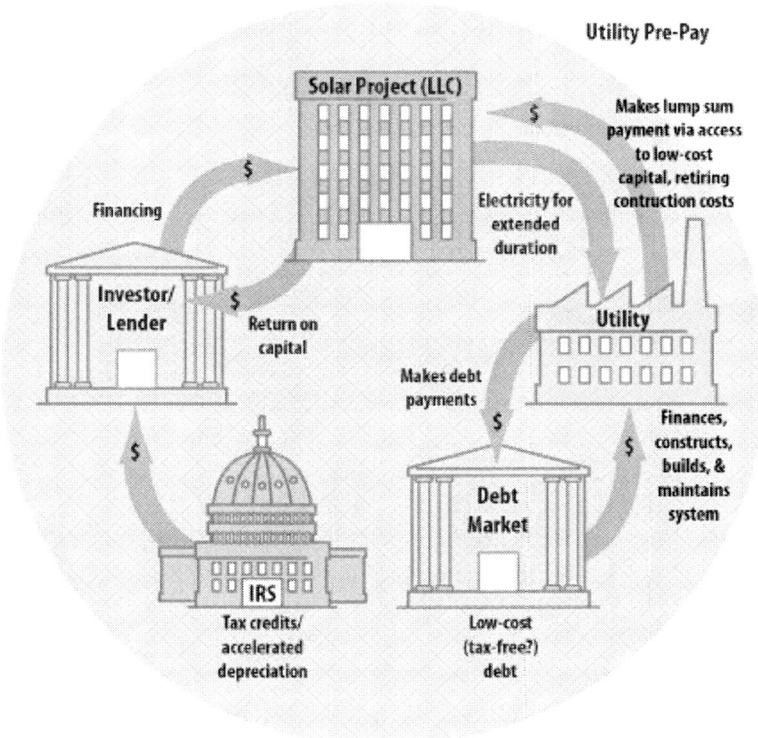

Figure 16. Utility prepay financial arrangement.

The deal also had an early buyout option to enable LADWP, along with Burbank and Pasadena, to acquire ownership of the facility after 10 years. SCPPA contracted with UPC Wind for the 20-year term, provided the project financing once the project was completed, and signed power sales agreements with LADWP and the other participants for the output of the project at a price sufficient to retire the debt and pay for the ongoing operating expenses.

In a similar arrangement, the 205 MW White Creek Wind Project in Klickitat County, Washington, was developed to take advantage of public power's access to tax-free debt. The project was initiated by four Washington utilities and became a reality with the help of a power prepay arrangement that accessed tax credits from the project and tax-free financing that reduced the cost of delivered power (Delony 2008). To finance the project, the utilities floated bonds on their balance sheets and prepaid for the power to be delivered under the project (Delony 2008).

While the aforementioned projects were related to wind, the utility prepay financial structure could benefit solar projects, and the structure takes advantage of low-cost municipal debt and federal tax credits associated with private project ownership.

CONCLUSION

Federal and state policies have provided crucial support for renewable energy project development. In response to these policies, developers of and investors in utility-scale solar projects have created innovative financial structures to maximize the benefits available and allocate risk and reward according to specific preferences. Each financing innovation has its own advantages and challenges and depends on various factors, including technology, counter-party, and the risk appetite of various investors.

Critical to successful utility-scale solar development, the federal ITC and accelerated depreciation programs provide a powerful set of incentives, representing approximately 50%–60% of a project's initial capital cost. In response to the 2008 financial crisis and difficulty in arranging specialized tax equity to monetize the ITC, the 1603 Treasury grant program provided additional stimulus. Through February 21, 2012, the 1603 program has awarded $10.76 billion to roughly 23,000 projects, stimulating an additional $25.11 billion in private market investment.

The DOE loan guarantee program also provided significant stimulus to the utility-scale solar sector. Section 1705 under the loan guarantee program supported $12 billion in loans, which will enable development of over 3,500 MW of solar generation capability. Section 1705 terminated on September 30, 2011, as required under its original statute.

At the state level, RPS requirements have spurred IOUs to sign long-term contracts with private developers or invest directly in renewable energy projects. At present, 29 states and D.C. have implemented RPS requirements.

To take full advantage of depreciation benefits and other tax incentives, the industry relies on complex financial arrangements, which include various forms of partnerships and lease structures. These financial arrangements are designed to allocate risk and reward among the developer and a specialized tax equity investor that can specifically take advantage of the various tax incentives made available at the federal and state level.

The perceived risk of utility-scale solar projects is still quite high and market acceptance remains elusive in this era of tight credit markets. For the immediate future, utility-scale solar development will remain dependent on the array of government incentives designed to support project financing. This will likely be true until solar technologies can provide energy at grid-parity (i.e., on par financially with competing technologies).

REFERENCES

Baker, Botts. (October 2007). "*Revenue Procedure Establishes a Safe Harbor for 'Partnership Flip' Structures in Wind Projects*." www.bakerbotts.com/file_upload/SafeHarborforPartnershipFlipStructuresinWindDeals.htm. Accessed June 25, 2010.

Barbose, G. (1 February 2010). *Email*. Electricity Markets and Policy Group, Lawrence Berkeley National Laboratory, Berkeley, CA.

Bartlett, J. & Goldstein, A. (19 August 2010). "*Q2 2010 Solar Industry Update*." Washington, DC: DOE.

Black & Veatch. (September 2007). *Arizona Renewable Energy Assessment*. B&V Project Number 145888. http://www.bv.com/Downloads/Resources/Brochures/rsrc_ENR_AZ_RenewableEnergy Assessment.pdf. Accessed January 27, 2010.

Bloomberg New Energy Finance. (2010a). "Blue Wing San Antonio PV Project." www.newenergyfinance.com/AssetFinancing/10614. Accessed June 2010.

Bloomberg New Energy Finance. (2010b). "SCE to Buy 50 MW Solar Power from Recurrent." www.newenergyfinance.com/News/36545/. Accessed June 2010.

Bolinger, M. (2009a). *Financing Non-Residential Photovoltaic Projects: Options and Implications*. LBNL-1410E. Berkeley, CA: Lawrence Berkeley National Laboratory. http://eetd.lbl.gov/EA/EMP/reports/lbnl-1410e.pdf. Accessed March 16, 2010.

Bolinger, M., Wiser, R., Cory, K. & James, T. (March 2009). *PTC, ITC, or Cash Grant? An Analysis of the Choice Facing Renewable Power Projects in the United States*. LBNL-1642E and NREL/TP-6A2-45359. Berkeley,

CA: Lawrence Berkeley National Laboratory and Golden, CO: National Renewable Energy Laboratory. http://eetd.lbl.gov/ea/emp/reports/lbnl-1642e.pdf. Accessed June 25, 2010.

Bracewell & Giuliani. (2009). "*Federal Stimulus Funding Fuels Further DOE Loan Guarantee Program Opportunities*." www.bracewellgiuliani.com/index.cfm/fa/news.advisory/item/8d731a90-b816-4993-833d-986b29e87fbc/Federal_Stimulus_Funding_Fuels_Further_DOE_Loan_Guarantee_Progra m_Opportunities.cfm. Accessed November 3, 2011.

California Public Utilities Commission (CPUC). (2009). "*Southern California Edison's Solar Photovoltaic (PV) Program*." http://www.cpuc.ca.gov/PUC/energy/Renewables/SCE+Solar+PV+Program.htm. Accessed January 11, 2010.

Carnahan, B. (15 February 2007). "*SCPPA Approves 200 MW Wind Purchase*." Southern California Public Power Authority. www.scppa.org/Downloads/Press%20Releases/SCPPAMilfordPressRelease20060215.pdf. Accessed March 17, 2010.

Chadbourne & Parke, LLP. (April 2010a). "*Strategies for Starting Construction*." Project Finance NewsWire. www.chadbourne.com/files/Publication/e68b24ed-1c27-4de3-b6e8-060e89217f7e/Presentation/PublicationAttachment/f6108758-cef5-44c8-84aa-0815e3eb2aa0/pfn0410.pdf. Accessed June 28, 2010.

Chadbourne & Parke, L. L. P. (February 2010b). "*The Year Ahead: What to Expect from Washington*." Project Finance NewsWire. www.chadbourne.com/files/Publication/0dc33c9f-254d-4acf-b869-d12778ec803a/Presentation/PublicationAttachment/e08eb620-e22d-4cfb-b258- d8e0f2e305cd/pfn0210.pdf. Accessed June 28, 2010.

Clean Energy States Alliance (CESA). (January 2009). "*State Clean Energy Fund Support for Renewable Energy Projects: Key Findings from the CESA National Database*." www.cleanenergystates.org/Publications/cesa-database_summary_v8.pdf. Accessed March 10, 2010.

Competitive Energy Insight, Inc. (2011). "*Summary of Lease Transaction Options*." www.ceinsight.com/blog/summary-of-lease-transaction-options. Accessed December 10, 2011.

Cooney, K., Winka, M., Freeman, R., Wobus, N. & Kallock, B. (January 2008). "The Cost of New Jersey's Solar PV Transition: An Analysis of Ratepayer Impacts Associated with Alternative Models for Transitioning a Statewide Solar PV Program from Rebates to Market-Based Incentives." Boulder, CO: Summit Blue Consulting.

Cory, K. S. & Swezey, B. G. (December 2007). *Renewable Portfolio Standards in the States: Balancing Goals and Implementation Strategies*. Golden, CO: National Renewable Energy Laboratory. www.nrel.gov/docs/fy08osti/41409.pdf. Accessed June 28, 2010.

Cory, K., Coughlin, J. & Coggeshall, C. (May 2008). *Solar Photovoltaic Financing: Deployment on Public Property by State and Local Governments*. NREL/TP-670-43115. Golden, CO: National Renewable Energy Laboratory. www.nrel.gov/docs/fy08osti/43115.pdf. Accessed March 12, 2012.

Cory, K., Coughlin, J., Coggeshall, C., Levite, B. & Kreycik, C. (2009). *Solar Photovoltaic Financing: Deployment by Federal Government Agencies*. NREL/TP-6A2-46397. Golden, CO: National Renewable Energy Laboratory.

Coughlin, J. & Cory, K. (March 2009). *Solar Photovoltaic Financing: Residential Sector Deployment*. NREL/TP-6A2-44853. Golden, CO: National Renewable Energy Laboratory.

Cuddy, S. (March 2011). "*A Look at the Cleantech Industry*." Moody, Famiglietti, and Andronico. www.mfa-cpa.com/alerts-and-insights/insights/2011/cleantech-perspective. Accessed February 22, 2012.

Database of State Incentives for Renewables & Efficiency (DSIRE). (2011a). "Massachusetts Incentives/Policies for Renewables & Efficiency: Renewable Portfolio Standard." www.dsireusa.org/incentives/incentive.cfm?Incentive_Code=MA05R&re=1&ee=1. Accessed December 1, 2011.

DSIRE (2011b). "*Colorado Incentives/Policies for Renewables & Efficiency: Renewable Energy Standard*." http://dsireusa.org/incentives/incentive.cfm?Incentive_Code=CO24R&re=1&ee=1. Accessed December 1, 2011.

DSIRE. (2011c). "*New Jersey Incentives/Policies for Renewables & Efficiency*." http://www.dsireusa.org/incentives/incentive.cfm?Incentive_Code=NJ05R&re=1&ee=1. Accessed December 1, 2011.

DSIRE. (2011d). "*Ohio Incentives/Policies for Renewables & Efficiency*." http://www.dsireusa.org/incentives/incentive.cfm?Incentive_Code=OH14R&re=1&ee=1. Accessed December 1, 2011.

DSIRE. (2011e). "*District of Columbia Incentives/Policies for Renewables & Efficiency*." http://www.dsireusa.org/incentives/incentive.cfm?Incentive_Code=DC04R&re=1&ee=1. Accessed December 1, 2011.

DSIRE. (2012a). "*Summary Maps: RPS Policies*." http://dsireusa.org/documents/summarymaps/RPS_map.pdf. Accessed March 2, 2012.

DSIRE. (2012b). "*Summary Maps: RPS Policies with Solar/DG Provisions.*" http://dsireusa.org/documents/summarymaps/Solar_DG_RPS_map.pdf. Accessed February 29, 2012.

DSIRE. (2012c). "*Summary Maps: State Tax Credits for Solar Proejcts.*" http://dsireusa.org/userfiles/image/summarymaps/solartaxcreditmap.gif. Accessed February 29, 2012.

DSIRE. (2012d). "*Summary Maps: State Property Tax Incentives for Solar Projects.*" http://dsireusa.org/userfiles/image/summarymaps/solarproperty taxmap.gif. Accessed February 29, 2012.

DSIRE. (2012e). "*Summary Maps: State Sales Tax Incentives for Solar Projects.*" http://dsireusa.org/userfiles/image/summarymaps/solartax creditmap.gif. Accessed March 2, 2012.

DSIRE. (2012f). "*Summary Maps: Public Benefit Funds for Renewables.*" http://dsireusa.org/documents/summarymaps/PBF_Map.pdf.Accessed February 29, 2012.

Delony, J. (January 2008). "*Utilities Turn To a 'Power Prepay' For White Creek.*" North American Windpower. www.nawindpower.com/e107_plugins/content/content.php?content.1730. Accessed March 17, 2010.

Department of Energy (DOE. (2009). "*Suggestions for Strong Loan Guarantee Applications.*" www.lgprogram.energy.gov/SS-LG-Apps.pdf. Accessed June 2, 2009.

DOE. (2010). "*DOE Announces Nearly $1.4 Billion in Conditional Loan Guarantees for BrightSource Energy.*" www.lgprogram.energy.gov/press/022210.pdf. Accessed September 15, 2011.

DOE. (2011). "*Loan Programs Office: Programs.*" https://lpo.energy.gov/?page_id=37. Accessed November 15, 2011.

Department of Treasury. (2012). "*1603 Program: Payments for Specified Energy Property in Lieu of Tax Credits.*" http://www.treasury.gov/initiatives/recovery/Pages/1603.aspx. Accessed January 20, 2012.

Doris, E., McLaren, J., Healey. V. & Hockett, S. (2009). *State of the States 2009: Renewable Energy Development and the Role of Policy*. NREL/TP-6A2-46667. Golden, CO: National Renewable Energy Laboratory.

Duval & Stachenfeld. (2008). "*For Solar and Wind Alternatives, Christmas Comes Early.*" Duval & Stachenfeld, LLP. http://dsllp.com/pdfs/Project%20Greenstorm.%20vII%20Oct%202008.pdf. Accessed March 2, 2012.

Engblom, A. (November 2011). "*UPDATE: NRG's Crane: Falling Cost of Solar Could Revolutionize Hub-and-Spoke Industry Model.*" SNL Alerts.

Feo, E. & Tracy, S. (2009). "*Commercial Finance: The Dark Arts of Leverage, Tax Equity, Leases and More.*" Presented at Solar Power International

2009. www.novoco.com/energy/resource_files/reports/sbt_finalpreso_102609.pdf. Accessed June 28, 2010.

GE Capital. (2010). "*Frequently Asked Questions*." GE Capital Energy Services Group. www.gecapitalenergy.com/quesb.html#one. Accessed March 10, 2010.

Haase, S., Billman, L. & Gelman, R. (January 2012). "*Western Region Renewable Energy Markets: Implications for the Bureau of Land Management*." NREL/TP-6A20-53540. Golden, CO: National Renewable Energy Laboratory.

Harper, J. P., Karcher, M. D. & Bolinger, M. (September 2007). "*Wind Project Financing Structures: A Review & Comparative Analysis*." Berkeley, CA: Lawrence Berkeley National Laboratory. http://eetd.lbl.gov/ea/emp/reports/63434.pdf. Accessed June 28, 2010.

Hastings, P. (July 2010). "*Additional Renewable Energy Cash Grant Guidance Released Regarding When Construction Begins*." StayCurrent: A Client Alert from Paul Hastings. www.paulhastings.com/assets/publications/1642.pdf. Accessed December 7, 2011.

Heeter, J. & Bird, L. (October 2011). *Status and Trends in U.S. Compliance and Voluntary Renewable Energy Certificate Markets (2010 Data)*. NREL/TP-6A20-52925. Golden, CO: National Renewable Energy Laboratory.

Holt, E. & Bird, L. (January 2005). *Emerging Markets for Renewable Energy Certificates: Opportunities and Challenges*. NREL/TP-620-37388. Golden, CO: National Renewable Energy Laboratory. www.nrel.gov/docs/fy05osti/37388.pdf. Accessed June 28, 2010.

Holt, E. A. & Wiser, R. H. (April 2007). *The Treatment of Renewable Energy Certificates, Emissions Allowances, and Green Power Programs in State Renewables Portfolio Standards*. LBNL-62574. Berkeley, CA: Lawrence Berkeley National Laboratory. http://eetd.lbl.gov/ea/ems/reports/62574.pdf. Accessed June 28, 2010.

Iberdrola. (8 November 2011). "*Congressman Grijalva and Energy Leaders Celebrate Arizona's Newest Solar Project at Dedication Event in Florence*."

Jacobs, R. (2009). "*Sale Lease-Back Financing: What Is It? How Can It Help Your Business?*" CashXchange Group. http://ukcreditcards.info/finance/sale-lease-back-financing-what-is-it-how-can-it-help-your-business-1073. Accessed March 17, 2010.

Jaffe, S. (4 August 2009). "*Robbing Renewable Energy to Pay for Clunkers.*" Greentech Media. www.greentechmedia.com/articles/read/robbing-renewable-energy-to-pay-for-clunkers/. Accessed March 17, 2010.

Kendall. (24 April 2009.) "'*Paying for Green Government': Leveraging 3rd Party Equity.*" www.efc.unc.edu/training/2009/GreenGov/Kendall.pdf. Accessed June 28, 2010.

Kimber, S. (January and May 2010). *Phone conferences. Recurrent Energy*, San Francisco, CA.

Klepper. (27 October 2009). "*Financing Opportunities Under the Department of Energy's Loan Guarantee Program.*" Presented at the Solar Power International Conference.

Lantz, E. & Doris, E. (October 2009). *State Clean Energy Policies Analysis (SCEPA): State Tax Incentives.* NREL/TP-6A2-46567. Golden, CO: National Renewable Energy Laboratory. www.nrel.gov/docs/fy10osti/46567.pdf. Accessed June 28, 2011.

Los Angeles Department of Water and Power (LADWP). (15 February 2007). "Wind Power Agreement Breezes Ahead: Project Would Provide 185 Megawatts of Clean Energy to LA." www.portarthurrefinery.com/go/doc/1475/162074/. Accessed June 28, 2010.

MarketWatch (2012). "*MidAmerican Energy Holdings Company Announces Expansion in the Renewable Energy Market*", January 24, 2012, http://www.marketwatch.com/story/midamerican-energy-holdings-company-announces- expansion-in-the-renewable-energy-market-2012-01-24. Accessed March 12, 2012

Mendelsohn, M. (2010). "*A Chicken in Every Pot? Is There Enough Tax Equity to Maintain the RE Market?*" NREL. https://financere.nrel.gov/finance/content/chicken-every-pot-there-enough-tax-equity-sustain-re-market. Accessed February 22, 2012.

Mendelsohn, M. (2011). "*Looking Under the Hood: Some Perspective on the Loan Guarantee Program.*" NREL. https://financere.nrel.gov/finance/content/looking-under-hood-some-perspective-loan-guarantee-program. Accessed February 22, 2012.

Mendelsohn, M. & Harper, J. (forthcoming). "*Section 1603 Expiration: Implications for Renewable Energy Project Financing.*" NREL. Golden, CO: National Renewable Energy Laboratory.

Mendelsohn, M., Kreycik, C., Bird, L., Schwabe, P. & Cory, K. (March 2012a). *The Impact of Financial Structure on the Cost of Solar Energy.* TP-6A20-53086. Golden, CO: National Renewable Energy Laboratory. http://www.nrel.gov/docs/fy12osti/53086.pdf. Accessed April 9, 2012.

Mendelsohn, M., Lowder, T. & Canavan, B. (April 2012b). *Utility-Scale Solar Concentrating Solar Power and Photovoltaics Projects: A Technology and Market Overview*. TP-6A20-51137. Golden, CO: National Renewable Energy Laboratory. http://www.nrel.gov/docs/fy12osti/51137.pdf. Accessed April 9, 2012.

Metzger, E. (April 2008). "*Bottom Line on Renewable Energy Tax Credits*." World Resources Institute. www.wri.org/publication/bottom-line-series-renewable-energy-tax- credits. Accessed March 10, 2010.

Miller, A. (March 2009). "*U.S. Stimulus Bill - New Tax Regime Boosts Power and Utilities*." Ernst & Young, LLP. www.ey.com/Publication/vwLUAssets/US_stimulus_bill/$FILE/US%20stimulus%20bill.pdf. Accessed March 17, 2010.

MoneyNews. (15 March 2012). "Buffett Plans New Bond Offering After Oversubscribed Solar Deal." http://www.moneynews.com/Investing Analysis/Greece-Debt-Insurance/2012/03/01/id/431086. Accessed March 15, 2012.

Murray, T. (2010). "Iberdrola Firms Up Solar Plans; Inks Supply Deal With SunPower—UPDATE." *Green Energy Reporter*. http://greenenergy reporter.com/renewables/solar/iberdrola-firms-entry-solar-biz-inks-supply-deal-sunpower/. Accessed February 22, 2012.

Public Service Electric and Gas Company (PSE&G). (2009a). "*PSE&G 'Solar 4 All' Program*." www.pseg.com/solar4all/Attachments/Solar4allfact sheet.pdf. Accessed July 7, 2010.

PSE&G. (2009b). "*PSE&G 'Solar 4 All'* Program Approved." www.pseg.com/solar4all/Attachments/July2909Solar4allrelease.pdf. Accessed July 7, 2010.

PG&E. (20 January 2010a). "PG&E Corporation and SolarCity Announce $60 Million in Financing to Install Solar Power for Businesses and Homeowners."

PG&E. (21 June 2010b). "PG&E Corporation and SunRun Create $100 Million Home Solar Financing Fund."

Ray, J. & Rawley, K. (May 2007). "The Financial Advantages of the Sale Leaseback." Tririga Research.

Recharge. (12 January 2012). "Montana's Rim Rock Developer Obtains Project Financing."http://www.rechargenews.com/energy/wind/article 297575.ece?WT.mc_id=re chargenews_rss. Accessed March 12, 2012.

Recurrent Energy. (July 2008). "*Recurrent Energy Secures $75 Million to Expand Solar as a Service Offering*." www.recurrentenergy.com/docs/

press/recurrent/2008/080716%20Hudson%20Funding%20Press%20Release.pdf. Accessed June 28, 2010.

Renewable Energy Finance Tracking Initiative (REFTI). (2011). "Project Status and Information." https://financere.nrel.gov/finance/REFTI. Accessed October 2011.

Schwabe, P. (17 July 2011). "Rim Rock Agreement Reached." NREL. https://financere.nrel.gov/finance/content/rim-rock-wind-agreement-reached. Accessed February 22, 2012.

Schwabe, P., Cory, K. & Newcombe, J. (July 2009). *Renewable Energy Project Financing: Impacts of the Financial Crisis and Federal Legislation*. NREL/TP-6A2-44930. Golden, CO: National Renewable Energy Laboratory.

Solar Energy Industries Association. (2011). "*U.S. Solar Market Insight 2nd Quarter 2011: Executive Summary*." www.seia.org/galleries/pdf/SMI-Q2-2011-ES.pdf. Accessed October 31, 2011.

SunPower. (29 June 2009). "*Wells Fargo Team to Finance $100 Million in Solar Projects*." SunPower Public Relations Announcement.

Tracy, S. (February 2010). "*Passive Loss Issues in Connection with Solar Investments*." Novogradac & Company LLP. www.novoco.com/journal/2010/02/news_qa_retc_201002.php. Accessed March 10, 2010.

Wilson, Sonsini, Goodrich & Rosati (WSGR). (October 2009). "FIPP, FIPP, Hooray…? Analysis of and Commentary on DOE's Financial Institutional Partnership Program." www.wsgr.com/publications/pdfsearch/wsgralert_fipp.pdf. Accessed March 12, 2012.

Wiser, R. & Barbose, G. (April 2008). "*Renewables Portfolio Standards in the United States: A Status Report with Data through 2007*." Berkeley CA: Lawrence Berkeley National Laboratory. http://eetd.lbl.gov/ea/ems/reports/lbnl-154e.pdf. Accessed June 28, 2010.

W. P. Carey. (2010). "Sale-Leaseback Financing." http://www.wpcarey.com/Real-Estate-Financing-Solutions/Sale-Leasebacks-Financing.aspx. Accessed March 17, 2010.

End Notes

[1] On January 1, 2017, the ITC will revert back to 10% (Bolinger et al. 2009).

[2] Clawbacks are, in effect, a requirement to return previously distributed money or credited funds.

[3] IRS rules on utility normalization mean that the benefit of the tax credit cannot be passed on to the utility's customers in a time period shorter than the life of the facility (approximated by

the length of the PPA). This essentially limits utility customer benefit from the ITC program. For more information, please see https://financere.nrel.gov/finance/content/normal-accounting-rules-limit-utility-ownership-renewable- energy-projects.

[4] See http://www.irs.gov/businesses/small/article/0%2C%2Cid=146335%2C00.html.

[5] According to grants.gov, a federal grant is "an award of financial assistance from a federal agency to a recipient to carry out a public purpose of support or stimulation authorized by a law of the United States." In contrast, the 1603 Treasury grant is a payment in lieu of receiving the ITC.

[6] From the "List of Awards" spreadsheet made available on the Department of Treasury website. http://www.treasury.gov/initiatives/recovery/Pages/1603.aspx. Accessed February 7, 2012.

[7] Number of awards data cited from the "Status Overview" document, available on the Department of Treasury website, last updated October 31, 2011. The aforementioned "List of Awards" spreadsheet, while more recently updated, groups many smaller projects and thus does not accurately represent the number of projects awarded.

[8] For several reasons, the 1603 awarded dollars may not match 30% of total expenditures precisely. The 1603 award, like the ITC, is only applied to "eligible" investment, which can exclude certain ancillary costs such as fencing and roads.

[9] See NREL REFTI results for Q1 2010 at http://financere.nrel.gov/finance/REFTI, slide 27.

[10] The monetization values represent the weighted averages of bin ranges selected by 17 industry participants participating in a 2010 REFTI webinar. Monetization of the ITC also referenced the accelerated depreciation benefits, which continue to require tax equity investment or direct use by the project developer.

[11] Other agencies, such as the Export-Import Bank, the Overseas Private Investment Corporation, the Department of Defense R&D fund, and U.S. Department of Agriculture also offer loan programs.

[12] All but one 1705 loan guarantee for solar is considered utility scale. The lone exception is the $1.4 billion loan guarantee to Prologis, the largest owner of warehouses in the country, which plans to develop a large number of rooftop PV sites with the loan guarantee. Individual projects under the Prologis effort are not yet known, although the overall project is predicted to produce 752 MW at peak output in 28 states.

[13] RPS policies are sometimes called renewable energy standards or alternative energy standards. These terms are interchangeable.

[14] Iowa's requirement was actually in terms of capacity, at 105 MW.

[15] See Heeter and Bird 2011. http://www.nrel.gov/docs/fy12osti/52925.pdf.

[16] SREC trading is active, or anticipated to become active, in D.C., Delaware, Maryland, Massachusetts, New Jersey, New Hampshire, North Carolina, Ohio, and Pennsylvania where solar ACPs are high. For example, New Jersey's solar ACP was set at $711/MWh for compliance year 2008/2009 and declines 2.5% per year for at least the next eight years (DSIRE 2011c). Ohio's solar ACP was set at $450/MWh in 2009 and will decline by $50/MWh every other year until stabilizing at $50/MWh in 2024, whereas D.C.'s is set at $500/MWh for 2009 through 2018 (DSIRE 2011d; DSIRE 2011e). Massachusetts's recently passed solar set-aside established a $600/MWh solar ACP for 2010 (DSIRE 2011a).

[17] Lawrence Berkeley National Laboratory estimates may represent over 600 MW of solar PV and CSP in the two states, assuming a capacity factor of 24% for a central station PV plant with tracking and a capacity factor of 30% for a CSP plant (Barbose 2010).

[18] Based on the expected mix of supplying projects and capacity factors.

[19] Developers generally have very high discount rates due to the risk associated with project development.

[20] The REFTI process included developers, financiers, and others associated with PV projects 1 MW or larger and CSP projects.

[21] Financing-related barriers include PPA/creditworthiness of power purchaser, finding tax equity investor, raising capital, and accessing government incentive programs. Non-financing-

related barriers include project economics, technological hurdles, environmental permitting, transmission, other, and none.

[22] The accompanying report, *The Impact of Financial Structure on the Cost of Solar Energy* (Mendelsohn et al. 2012a), provides a description of and quantitative assessment on four specific structures: single owner, sale-leaseback, and two forms of partnership flip [all-equity and levered (i.e., with debt)].

[23] There is no equivalent requirement for independent developers.

In: Utility-Scale Solar Installations
Editor: Miles B. Simmons

ISBN: 978-1-62948-999-5
© 2014 Nova Science Publishers, Inc.

Chapter 3

THE IMPACT OF FINANCIAL STRUCTURE ON THE COST OF SOLAR ENERGY*

Michael Mendelsohn, Claire Kreycik, Lori Bird, Paul Schwabe and Karlynn Cory

EXECUTIVE SUMMARY

To stimulate investment in renewable energy generation projects, the federal government developed a series of support structures that reduce taxes for eligible investors—the investment tax credit, the production tax credit, and accelerated depreciation. The nature of these tax incentives often requires an outside investor and a complex financial arrangement to allocate risk and reward among the parties. These financial arrangements are generally categorized as "advanced financial structures." Among renewable energy technologies, advanced financial structures were first widely deployed by the wind industry and are now being explored by the solar industry to support significant scale up in project development.[1] This report describes four of the most prevalent financial structures used by the renewable sector and evaluates the impact of financial structure on energy costs for utility-scale solar projects that use photovoltaic and concentrating solar power technologies.

* This is an edited, reformatted and augmented version of the U.S. Department of Energy, Office of Energy Efficiency & Renewable Energy, NREL/TP-6A20-53086, dated March, 2012.

A critical aspect of the analysis is the input assumptions, including the cost to install each technology, the costs and terms of financial capital relevant to the technology and financial structure assessed, and the operating costs of each technology. To determine reasonable inputs, the authors conducted a series of interviews with industry experts, reviewed several sources in the current literature, and relied on the default values from the System Advisor Model that was employed in this analysis, which were, in turn, guided by the industry expertise of its developers.

The analysis determined that financial structures that include project-level debt generally yield a lower levelized cost of energy (LCOE) compared to those that rely purely on equity capital, although in practice raising debt at the project level can be difficult, particularly for developers without sizable balance sheets and a strong history of development experience. Other insights from the analysis include:

- Debt associated with the loan guarantee program can reduce LCOE by approximately 20%, and possibly more, depending on the quantity of debt the project is allowed to take on.
- Certain financial experts expect "tax equity" capital—incorporated to maximize the use of federal tax credits and depreciation benefits—to increase in cost by 200 to 400 basis points (bp) (or 2%–4%) due to the termination of the current 1603 Treasury grant program at the end of 2011.[2] Such an increase in the cost of tax equity is projected to raise the LCOE from utility-scale solar projects by 3%–20%.
- The cost of tax equity is particularly critical in all-equity financial structures and can have a significant impact on the cost of energy produced. Even more important in all- equity financial structures, the internal rate of return (IRR) target year—the project year in which the tax equity investor is expected to earn the pre-negotiated return— can have a very large impact on the resulting cost of energy. Delaying the IRR target year, for example, from year eight of the project to year nine, can improve the LCOE by 7%–27%, depending on the technology and financial structure employed.
- Development and operational experience is projected to lead to further financial structure innovation and reduction in required investment returns for various sources of financial capital.

1. INTRODUCTION

The U.S. government has established several mechanisms to support investment in renewable energy generation facilities through the tax code. The two most critical support structures offered by the federal government are (1) investment or production tax credits (ITC or PTC), and (2) an accelerated depreciation schedule for renewable energy projects known as the Modified Accelerated Cost Recovery System (MACRS) and a "bonus" depreciation that allows for additional acceleration of the tax benefits to investors.[3] Both the ITC/PTC and MACRS mechanisms can benefit the project by reducing the taxes payable, thereby improving the economic performance of the project. In total, the economic benefit of the federal tax incentives can represent roughly 50%–60% of the installed cost of the solar project (Bolinger 2009). In recent years, there has also been a cash grant available through the Treasury 1603 program in lieu of the ITC, which expired at the end of 2011.

However, renewable energy project developers typically do not have sufficient taxable income to take full advantage of the tax incentives directly.[4] Accordingly, renewable energy project developers have implemented a wide array of complex financing structures with specialized "tax equity investors" (typically large investment banks or insurance companies) that have sufficient taxable income from other business activities and expertise to take advantage of the tax benefits.

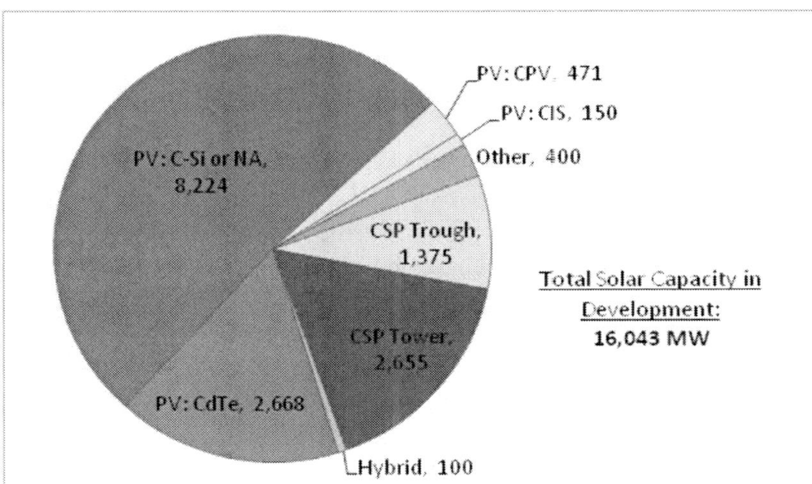

Figure 1. U.S. utility-scale solar projects currently under development.

These financial structures, such as "partnership flips" and various lease structures, have been instrumental in the maturation of the wind industry. As the photovoltaic (PV) and concentrating solar power (CSP) industries continue to develop utility-scale facilities, advanced financial structures are likely to play an increasingly important role in the allocation of risk and reward among different investor classes.

According to the National Renewable Energy Laboratory's (NREL's) database of utility-scale projects of 5 megawatts (MW) or greater, more than 16,000 MW are currently under development and hold a power purchase agreement (PPA) with a utility or are under development by a utility (see Figure 1).

According to a study by the Lawrence Berkeley National Laboratory (LBNL), the cost of wind energy can be highly influenced by the financing structure utilized, ranging from $48 per megawatt-hour (MWh) to $63/MWh for a hypothetical project (Harper et al. 2007). The authors found the variation in costs to result primarily from (1) financing-related transaction costs, (2) assumptions about internal rate of return for tax and equity investors, and (3) the relative terms of each structure including the level of equity versus debt. Factors that were found to be important in selecting financing structures were: simplicity, standardization, and speed of the financing process. While financial structures utilizing project-level debt were determined to have a lower cost of energy in analytical models, the use of debt was perceived by actual equity investors as expensive to implement (i.e., requiring costly due diligence), complex, and leading to a loss of control (Harper et al. 2007).

The purpose of this analysis is to examine the effects of different financial structures on the cost of energy from PV and CSP plants [with differentiation between parabolic-trough CSP (CSP– Trough) and power-tower CSP (CSP–Tower) technologies]. The analysis relies on modeling results from the System Advisor Model (SAM),5 developed by NREL, in collaboration with Sandia National Laboratories (SNL) and in partnership with the U.S. Department of Energy (DOE) Solar Energy Technologies Program. In early 2011, NREL added the capability to assess complex financial structures to SAM (details provided in text box on page 17).

This study examines the levelized cost of energy (LCOE) impact due to financial structure and terms for CSP and PV technologies. The report first describes in detail four commonly used financial structures—single owner, all-equity partnership flip, leveraged partnership flip, and sale leaseback—and their use in the solar industry. Next it presents results from the SAM model comparing the LCOE for CSP and PV technologies under the four different financing structures. The assumptions used in the modeling are informed by interviews with renewable energy industry financing experts and relevant literature. The study then examines the sensitivity of the LCOE to changes in key financial variables, including tax equity returns, interest rates, and the duration of partnership flips. Finally, conclusions are offered based on the modeling results and interviews with financial experts.

This report is one in a series of three reports on utility-scale solar installation in the United States. The other reports provide (1) a primer on utility-scale solar technologies and market activities, and (2) an overview of the policies that support financing of utility-scale solar systems.

2. Background on Financing Structures

This analysis focuses on four commonly used financial structures: single owner, all-equity partnership flip, leveraged partnership flip, and sale leaseback. Innumerable variations on these structures can be applied depending on the risk appetites and reward expectations of the parties involved, including allocation of tax and cash benefits, buyout provisions during set periods along the project life, and default allocation of risk to parties. The following background information on these structures derives primarily from two sources: *Wind Project Financing Structures: A Review and Comparative Analysis* (Harper et al. 2007) and *Renewable Energy Project Financing: Impacts of the Financial Crisis and Federal Legislation* (Schwabe et al. 2009). Other sources are cited separately.

The key elements of the four financing structures are summarized below and in Table 1.

2.1. Single Owner

The single owner structure involves one project owner, which can be either the project sponsor (developer) or a large taxable entity that has purchased the project from a developer. The single owner receives all of the cash and tax benefits (see Figure 2).7 Generally, an entity that enters into a single-owner structure for a renewable energy generation project has the ability to use all of the tax benefits from the project.[8]

Table 1. Summary of Key Elements of Financing Structures (as modeled for analysis)[6]

	Single Owner	All-Equity Partnership Flip	Leveraged Partnership Flip	Sale Leaseback
Equity Owners	Developer (third party if sold)	Tax investor; developer	Tax investor; developer	Tax investor (lessor)
Project Debt	Yes (owner choice)	None	Yes	None
Return Target	Owner After- Tax IRR	Tax investor after- tax IRR (flip target)	Tax investor after-tax IRR (Flip Target)	Lessor after-tax IRR
Cash Sharing	Owner: 100% of project cash	Pre-Flip: all to tax investor after developer cost recovery; Post-Flip: primarily to developer	Pre-Flip: Proportional to investment after developer cost recovery; Post-Flip: Primarily Developer	Lessor: Receives lease payment; Lessee: Receives operating budget, project margin
Tax Benefit Sharing	Owner: 100% of project tax benefits	Pre-Flip: primarily tax investor; Post-Flip: primarily developer	Pre-Flip: primarily tax investor; Post-Flip: primarily developer	Lessor and lessee different taxable incomes; ITC & depreciation to lessor

Sources: Karcher et al. 2010; SAM User Guide 2011.

The single owner may or may not utilize project-level debt to finance the facility. The analysis contained herein assumes project-level debt was employed for the single-owner structure.

The developer or investor often creates a special purpose entity for the project assets and pays the project's capital costs with funds from other operations. All of the project cash flows, tax incentives, and depreciation benefits remain with the owner. This structure is widely used for renewable energy projects and represents one of the simplest methods of owning and operating a project. There is no third-party equity requirement and there are no flip targets. The owner may acquire project-level debt or obtain financing at the company or parent level. The original SAM "independent power producer" (IPP) financial model follows this structure.

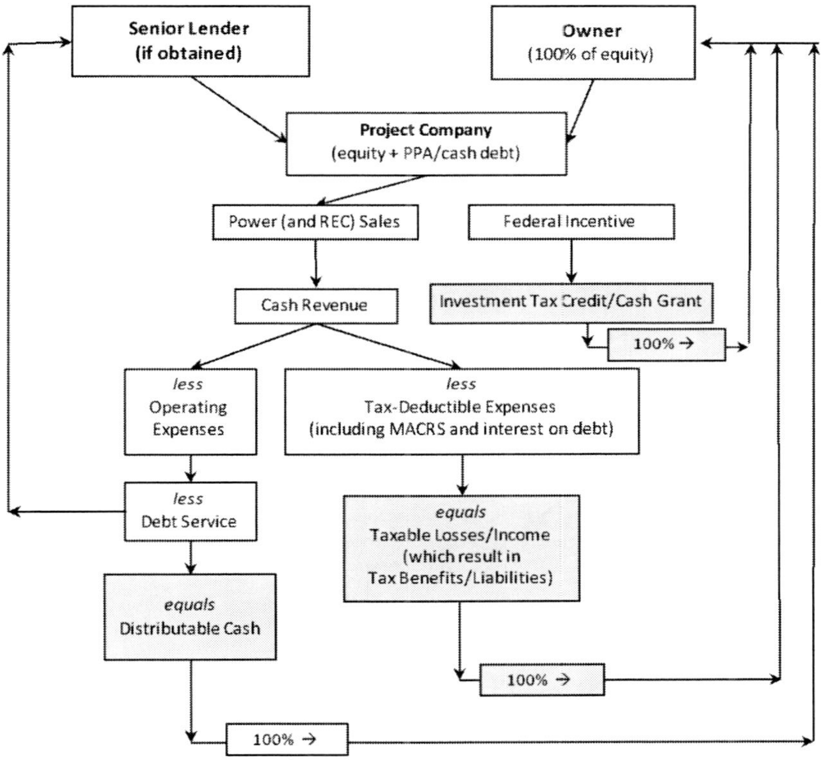

Source: Harper et al. 2007.

Figure 2. Single owner example structure.

2.2. All-Equity Partnership Flip

The all-equity partnership flip structure (also known as the pre-tax, after-tax partnership structure) involves two classes of investors in the partnership: a tax equity investor and a developer, also known as the project sponsor (SAM User Guider 2011). As the project structure name implies, all capital is provided as equity; there is no debt at the project level (although the investors can hold leverage on their corporate balance sheets).

Generally, an all-equity partnership flip is structured so the tax investor contributes a significant fraction of the required equity (e.g., 45%–65%)9 (see Figure 3). The tax equity contribution negotiated between the parties is often a function of the tax equity return requirement and the target year (Karcher et al.

2011). That is, a lower contribution from the tax equity investor will generally lead to higher returns on his investment, and vice versa. The remaining capital comes from the developer.

The tax benefits include the ITCs and the taxable losses, which may or may not be allocated similarly. Taxable losses—if there are any—are calculated as the cash revenue less all deductible expenses, including the accelerated depreciation benefits described above.

The project generates cash revenue through the sale of energy and renewable energy certificates (RECs), or possibly other revenue sources such as steam. Cash revenue less the payment of direct-operating expenses equals what is referred to as "distributable cash" available to the investors.

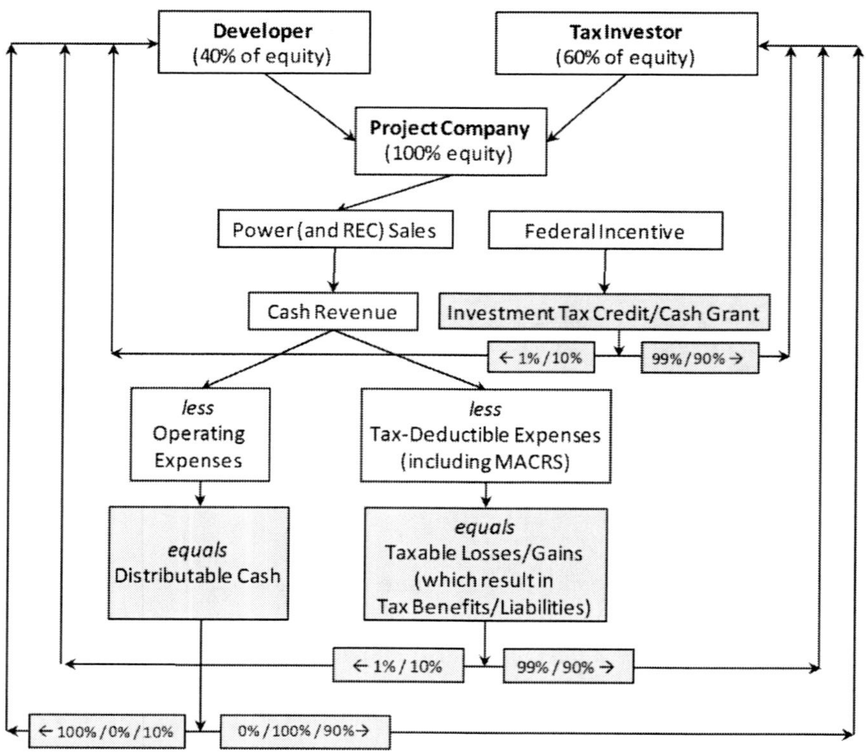

Source: Adapted from Karcher et al. 2010.

Figure 3. All-equity partnership flip example structure and allocations.

Initially, before a pre-designated "flip-point," the tax investor receives the vast majority (i.e., up to 99%) of the tax benefits from the ITC and taxable losses (see Figure 3) (the flip-point is designated as the "/" symbol).[10] Under Internal Revenue Service (IRS) guidelines, at least 1% of the tax benefits must be allocated to the developer in a partnership arrangement.[11] The tax investor may also receive some of the cash benefits associated with the sale of power and RECs in order to allow for the pre-negotiated return-on-tax equity.

The all-equity partnership flip generally treats the distribution of cash (equal to energy and REC revenues less operating expenses) differently than the allocation of tax benefits. For example, under one typical variant (shown in Figure 3), the developer gets 100% of the cash until he has recovered his initial capital investment.[12] Developer profit is essentially treated in a separate "development fee" in the SAM model. This fee is set at 3% of the total project cost for this analysis and as the default value in SAM.

After the developer recovers its initial investment (designated as the first slash in the bottom boxes of Figure 3), the developer receives 1% of the distributable cash. The tax equity investor receives 99%. After the flip point is reached and when the tax equity investor's return hits the pre-negotiated target (designated as the second slash in the same boxes), the allocation of cash will change again. This analysis assumes the developer gets 10% of the distributable cash and the tax equity investor 90% after the flip point, also the default values in SAM.

The project will reach the designated flip point once the total return (from tax and cash benefits) earned by the tax equity investor reaches a pre-negotiated internal rate of return (IRR). After the flip point is reached, the benefit streams are reallocated among the investors according to the pre-designed arrangement. The flip point is often referred to as the IRR "target year."

Typically, the allocations and equity contributions are designed so that the flip point occurs after the tax benefits have been fully realized and the five-year ITC "recapture" period has expired. During the recapture period, the IRS requires investors that utilize tax benefits to hold onto the assets or risk losing all tax benefits enjoyed since commercial operation began.

Some all-equity partnership flip structures utilize more flip points and associated reallocations of project benefits than represented in Figure 3, depending on the desires of the investors.[13]

2.3. Leveraged Partnership Flip

The leveraged partnership-flip structure is similar to the all-equity partnership flip but also includes project-level debt. The fraction of debt is determined by the debt terms, the cash generated by the project, and the limits placed by the lending institution. The loan is often sized such that it can be repaid from project cash flow and secured by the project's assets. The SAM model calculates the fraction of debt as the maximum allowable level given a user-defined debt- service coverage ratio (DSCR). The DSCR represents the additional cash available to service the debt.

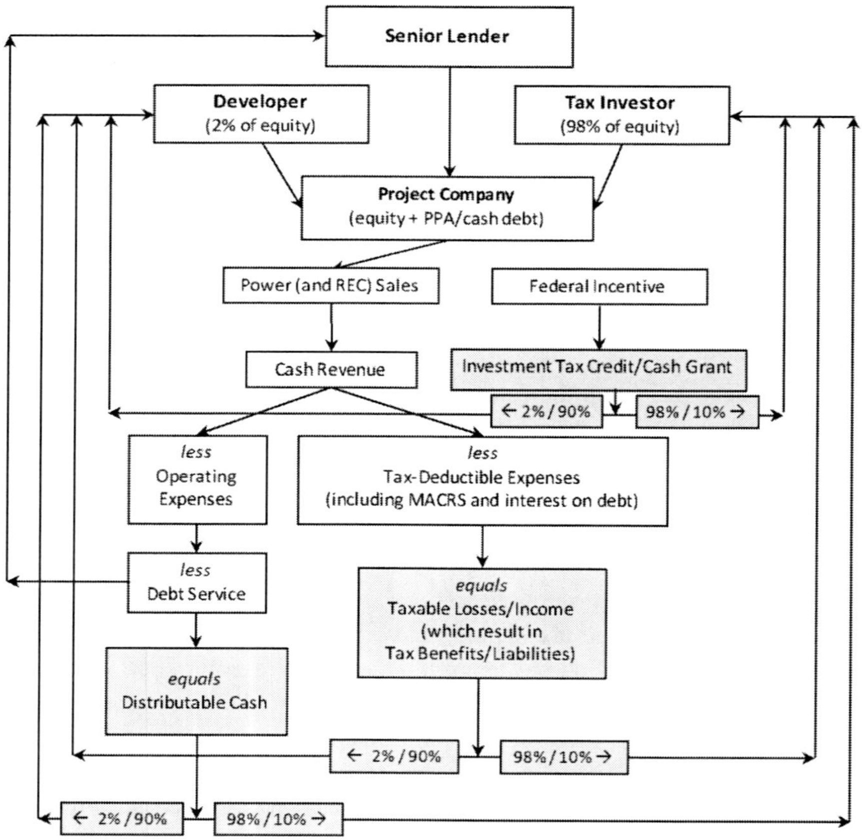

Source: Adapted from Harper et al. 2010.

Figure 4. Leveraged partnership-flip example structure and allocations.

Typically, in a leveraged partnership flip, the tax investor contributes virtually all of the equity required after the debt is secured and receives a pro rata allocation (i.e., proportional to its equity contribution) of both cash and tax benefits until it has achieved the pre-negotiated flip target. The developer generally makes a very small equity investment (e.g., 2% of the total equity invested) (see Figure 4).

Like the all-equity partnership flip, the benefit allocations within the leveraged partnership flip are usually designed to provide a flip point within years six to nine of the solar project's life. After the flip point is reached, the benefits are reallocated to provide the project developer the vast majority (e.g., 90%) of the project cash and remaining tax benefits. As with the equity flip, the post-flip allocations are determined to allow the tax investor to achieve a specific 20-year IRR. The equity contributions and allocation fractions can vary substantially by project. Because the lenders have the first lien on project assets, the equity IRR requirements are higher than with an all-equity partnership flip.

2.4. Sale Leaseback

The sale leaseback structure involves a tax equity investor who purchases the project from the developer and then leases it back (SAM User Guide 2011). In this structure, the tax investor (the lessor) leases the facility back to the developer (the lessee) and receives lease payments. The tax investor holds all of the tax credits and depreciation benefits from the project (see Figure 5).

The lessee operates the project and receives revenues from sales of electricity. The lessee also typically retains excess cash flow after operating costs and lease payments are made, which provides an incentive for the developer to operate the project efficiently.

Unlike the flip structures, under the sale leaseback structure, the tax equity investor and developer each have their own taxable income and project cash flows; there is no revenue sharing. However, there are cash flows being exchanged between the two parties as a result of the lease arrangement. The return target is the tax investor's IRR requirement at a given point in time, typically the end of the lease term.

Usually, the purchase price is based on the underlying project value relative to the tax-investor- required rate of return and known revenues, such as a signed PPA. The lease payments are sized based on a pre-defined lease coverage ratio to ensure there is adequate cash in the project to cover the lease payments. The purchase price of the project is based on the present value of lease payments, tax incentives, and depreciation, and is usually supported by an appraisal.

3. Selection of Financing Structures in the Marketplace

The authors conducted a series of interviews with the following industry experts selected for their knowledge of financial structures and large-scale solar development: Matthew Meares, Amonix; Matt Ferguson, Reznick Group; Joshua Turner and Jordan Roberts, BrightSource Energy; Daniel Englander, Agile Energy; John Harper, Birch Tree Capital; Adam Kobos, Stoel Rives; and Mike Niver, SolarCity. Interview topics ranged from the cost and availability of various sources of capital to pros and cons of specific financial structures.

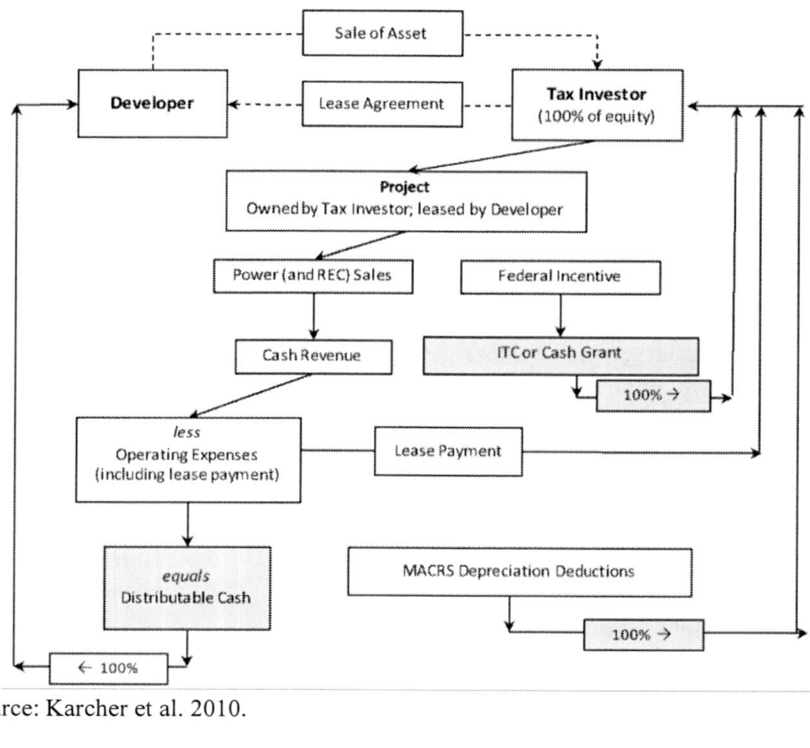

Source: Karcher et al. 2010.

Figure 5. Sale leaseback example structure.

The selection of financing structures for any particular project is based on a variety of factors. Expert interviews suggest that these factors include: project size, experience and comfort with the structure, the strength of the developer's balance sheet, the ability to use tax credits, and the investor's risk tolerance and preferences. Harper et al. (2007) also found simplicity, standardization, and speed of the financing process to be important in selecting financing structures.

Interviews also indicated that, because tax equity is in such high demand by a wide array of projects both inside and outside the renewable energy industry, the tax equity investor generally dictates many of the terms of the agreement including the financial structure selected. The tax equity market was often categorized by interview candidates as quite constrained and difficult to access, particularly for projects proposed by newer developers or those employing technologies without significant operational experience.

Tax equity investors frequently specialize in a particular structure due to experience and general comfort with the associated risks. Interviews with developers indicate that they will utilize a variety of structures, the final selection generally depending on tax investor preference and the expected risk/return tradeoff of the offers. The investors in these projects may regard the tax credit incentives—ITC, PTC, and cash grant—differently due to their variable impacts on LCOE and return expectations. Whether the incentive is production-based or investment-based impacts the present value of cash flow to equity. The tax equity investor may have the final say in which financing structure is used.

All interviewees advised developers to evaluate their projects under a range of potential financial structures before setting the PPA with the utility or other off-taker. The reality of the financial structure required by the tax investor may force the PPA to be framed in a certain way or cause higher expenses than assumed by developers when setting the PPA.

Lease structures or master lease agreements, which can be used to streamline a large number of projects, are common for PV in particular. One developer indicated he does not use flip structures because he does not have the tax appetite to take even a small amount of tax credits, while another indicated that some investors are not able to enter into partnership deals because of limitations in their charters. Flip structures can also be challenging for smaller utility-scale projects because of the cost and complexity of structuring the partnership. Also, there is greater uncertainty in the timing of payment with flip structures than with lease structures. Another developer indicated that flip structures are the most forgiving and the easiest to administer, with the fewest number of steps to undertake. The choice of structure can also be influenced by the developer's interest in monetizing cash flows. Accounting treatment of revenues and losses can be vastly different between structures, which may result in the developer or equity investor preferring one structure or another.

One interviewee in the PV industry indicated that, presently, it is difficult to negotiate debt when tax equity investors are involved, due to issues of forbearance and ownership.[14] Forbearance is a special agreement between the lender and the borrower to delay a foreclosure. While the tax equity investor is the owner of the project, senior debt would have a first lien on the assets in the case of bankruptcy. If a project defaults on its debt payments and the lender places a lien on the project, the owner may need to forfeit the tax incentives.

The details of the forbearance and ownership are very complicated, often requiring extensive legal review. For that reason, debt is generally only available at the project level to very large projects or a portfolio of projects. One interviewee indicated that debt lenders prefer to be engaged only in projects of $25 million to $50 million or larger. This is generally true whether the project incorporates direct ownership (i.e., single owner structure), a partnership flip, or a lease structure of some kind (referred to as a levered lease transaction). The only CSP-Trough facility completed of late (the Nevada Solar One Plant developed in 2007) utilized a levered lease transaction.

Joint ventures (also known as strategic partnerships) may be employed in some circumstances in lieu of a partnership flip. In contrast to a partnership flip, a joint venture represents a business venture with equity partners that take pro rata allocations of cash and tax benefits consistent with their investments in the joint venture and are not expected to exit the project after hitting a specified project return.[15] Interviewees indicated that joint ventures may be used when tax equity investors perceive that there is too much risk to invest or when a joint venture provides synergies or other benefits.

In one example of a recent joint venture, BrightSource Energy has raised equity at the corporate level and partnered with companies, including NRG Energy and Google, on the CSP-Tower project, Ivanpah Solar Electric Generating System (BrightSource 2011). However, the financial structure of the Ivanpah facility could still evolve as the parties weigh their options during or after construction. Overall, there is minimal market data to assess the preferred financial structure of the market to develop CSP facilities. As mentioned, only one CSP-Trough project was built in the recent past, and no CSP-Tower facilities have been completed recently in the United States.

Project structures frequently allow the developer, or sponsor, to re-purchase the project from the tax equity investor during various intervals along the project life. These repurchase options can provide significant value to the sponsor but are often based on qualitative inputs (e.g., "it's consistent with their business model") or require nuanced modeling beyond the capabilities of NREL's SAM model. Accordingly, the analysis in the following section ignored issues of repurchase or buyout.

4. Analysis and SAM Model Overview

This analysis examines the impact of financial structure on energy costs for large-scale solar PV and CSP plants, using the metric of real LCOE. Simply stated, LCOE is a discounted measure of the costs of building and maintaining an electric generating system divided by the discounted amount of energy produced over the analysis period. Both direct expenditures on the generating system and indirect costs, such as transaction costs, returns, and interest rates, contribute to the LCOE.

To complete this analysis, NREL employed the pro forma financial cash flow model that has been integrated into the SAM model to calculate the LCOE and other metrics under the four financial structures discussed above.[16] SAM allows users to investigate the impact on LCOE (among other output variables) of variations in project location and physical characteristics, technology performance and cost, and project financing parameters. The authors developed model input assumptions based on market intelligence gathered via the interview process, current literature on the topic, and the default values embedded in the SAM model (also developed by research and industry experts). See Financial Analysis Methodology below.

SAM's pro forma financial model includes a representative template for each of the project finance structures examined, which is a new feature not included in earlier versions of the model (see text box for differences between the new and old version of SAM).[17] For each of the Advanced Utility IPP financing structures, the model incorporates the assumed rates of return by the developer or tax equity investor, the equity contribution ratios, and the cash and tax benefits allocations, and then solves for the LCOE needed to satisfy those input assumptions.[18] The model also includes a wide array of input assumptions associated with power production, installed costs, and facility operations and maintenance (O&M).

5. Technology and Economic Assumptions

This analysis examines the LCOE of utility-scale CSP-Trough and CSP-Tower plants (100 MW nameplate capacity) and PV projects (20 MW-direct current nameplate capacity) using the four financial structures described above. The hypothetical plants are sited in Blythe, California (for CSP-Trough and CSP-Tower), and Phoenix, Arizona (for PV).

Table 2 summarizes the key cost and performance variables assumed in the analysis. These values are the default inputs for SAM at the time of the model runs, and are only intended to be generally representative of project costs. The PV cases assume $4.30/watt (W) fully installed ($86 million total) and incur fixed O&M expenses of $10/kW-year. The PV cases also assume single-axis tracking and a 78% direct current to alternating current derate factor. The analysis assumed CSP-Trough and CSP-Tower systems had installed costs of $7.50/W and $6.05/W, respectively. Fixed O&M costs for CSP technologies are assumed to be $70/kW-year and variable O&M costs are assumed to be $3/MWh.

For all cases, a 35% federal tax rate, 7% state tax rate, and a 5% sales tax was applied. Ongoing expenses were inflated at a rate of 2.5% per year. Annual insurance costs were set at 0.5% of initial capital costs. All cases also assumed a 30% ITC and MACRS depreciation with 50% bonus depreciation.

No other incentive structures, such as REC payments, were assumed available. All cases were modeled for 25 years and on the assumption that the developer does not exercise its option to purchase the investor's interest in the project.

Table 2. Power Plant Cost Assumptions

Primary Result	CSP-Trough	CSP-Tower	PV
Project Size(MW)	100	100	20
Installed Cost ($/watt)*	$7.50	$6.05	$4.30
Capacity Factor	41%	42%	26%
Generation (Annual MWh)	357,587	367,769	22,348
Fixed O&M	$70/kW-year	$70/kW-year	$10/kW-year
Variable O&M	$3/MWh	$3/MWh	-

* The installed cost is calculated as the "overnight" installed cost, and represents all fixed costs but excludes interest during construction and other financing costs.

Power production is calculated via SAM. All other performance variables are based on SAM model defaults (version 2011-06-30). A detailed description of the model's input assumptions and default values is provided in the SAM financials user guide (Gilman 2011).

SAM's Original IPP Model Versus New Advanced IPP Models

The new financial structures incorporated in SAM version 2011-06-30 are referred to as Advanced Utility IPP options. The prior versions of SAM included three IPP modules relevant to utility-scale projects, which are now rolled into a single Utility IPP case.

There are a number of differences between the original Utility IPP model and the new Advanced Utility IPP options. The Utility IPP model can be considered a simplified version of the current "Single Owner" model, which is included within the Advanced Utility IPP options. However, the two cases may not provide equivalent results for what appears to be the same project due to unique treatment of a few input variables.

The following elements are new to the Advanced Utility IPP models:

- Reserve accounts for debt service and working capital – Reserve accounts are generally required by lenders and investors to improve security of their investments and represent an important cost of financing.
- Sculpted debt – Allows specific year-to-year debt repayment based on the annual project cash flows. The quantity of debt, or debt fraction, is calculated as the maximum that can be loaned to the project given a user-defined constant debt service coverage ratio (DSCR). In contrast, the historic Utility IPP model allowed users to specify a debt fraction and a mortgage-style debt (equal annual payments).
- Developer fee – The Advanced IPP models allow the user to define a developer fee that can be paid upon project commercial operation or from cash flows over a specified time period. The original IPP model does not allow for payment of a development fee.
- MACRS depreciation schedules – The original Utility IPP model included the mid- quarter MACRS schedules.

Although the mid-quarter convention can apply to a solar facility depending on the facts and circumstances, mid-quarter schedules cannot be accurately modeled in an annual forecasting tool such as SAM. The feature was removed in the new version of SAM for all IPP cases. Mid-year MACRS schedules (shown in SAM as "5-yr MACRS") are now the default values and were applied for all cases in this analysis.

- Change in property tax accounting – The original Utility IPP model included default values that projected increasing real estate values and tax increases over time. The Advanced IPP model includes default inputs based on declining real estate values for tax purposes consistent with common property tax valuation practices.
- Financial closing costs – The Advanced Utility IPP models include cost components associated with the acquisition of financial capital, including equity-closing costs and debt-closing costs. The original Utility IPP model did not include these inputs.

6. Financial Analysis Methodology

To estimate the relative impacts of financing variables on the resulting cost of solar energy, both single variable sensitivities and scenarios were developed and compared to reference scenarios. Input values for reference and alternative scenarios were chosen based on multiple sources, including:

- Default values within the SAM Advanced Utility IPP models
- Interviews with a number of industry experts experienced with project development and underwriting
- 2010 Mintz Levin report, *Renewable Energy Project Finance in the U.S.: An Overview and Midterm Outlook,* which provides technology-specific debt and equity returns.

6.1. Reference Cases

Tables 3 and 4 present the financing variables used in the analysis for PV and CSP, respectively. One interviewee indicated that the interest rate might be 100 basis points higher for CSP-Tower systems because of the greater technology risk. Accordingly, this incremental return was incorporated in the assumed cost of debt and equity. Also, interviewees indicated that, except for small differences in DSCR, financial variables would not differ significantly for cadmium telluride thin film—the primary commercially available thin-film technology utilized in utility- scale installations at present—and crystalline PV technologies. The analysis did not consider financing costs for technologies not currently under development due to the necessary speculation of input assumptions.

For both PV and CSP, the partnership-flip structures were assumed to have an IRR target of year nine (i.e., the project's tax investors would reach their target IRRs in year nine of the project if the cash and tax benefits were realized as projected). Equity target returns for PV plants were assumed to be 9% and 11% for the all-equity and leveraged partnership flips, respectively. The equity returns for the leveraged partnership-flip structure are higher because the structure includes a creditor who retains first lien on the project's cash; the tax investor is assumed to require a higher yield because of this added risk.[19] Interviews and other resources indicate that those spreads are approximately two percentage points (200 basis points). For CSP technologies, the target returns are assumed to be higher due to technology risk given that only a small number of projects have been installed in the United States: 12% for all-equity flips and 14% for a leveraged flip (Mintz Levin 2010). However, interviewees indicated that there is no real market experience on which to base these rates.

For structures that allow for project-level debt, including the leveraged partnership flip and single-owner structures, key analysis inputs include the debt interest rate, term, and fees relevant to acquiring the debt instruments. The PV reference case debt rate is assumed to be 7% for 18 years, while the CSP debt rate modeled is 8% for 18 years. Other primary financial inputs to the structural analysis include the equity closing costs, development fees, and project benefit allocations.

The DSCR was modeled at 1.3 for all PV and CSP references; a level selected after consulting with industry experts. This means that cash flows must be 1.3 times the amount of debt owed per payment period. One interviewee suggested that, though CSP projects have higher technology risk, energy flows are quite stable. A debt margin for CSP technology risk can be built into the interest rate, reserve requirements, or the DSCR.[20]

Table 3. Key Financing Variables by Financial Structure – PV Reference Case

Financing Variable	All-Equity Partnership Flip	Leveraged Partnership Flip	Sale Leaseback	Single Owner
(Tax equity) IRR target year	9	9	20	20
(Tax equity) IRR target	9.0%	11.0%	9.0%	11.0%
Equity closing costs	$300,000	$300,000	$300,000	
Development fee	3%	3%	3%	
Tax investor contribution to equity	60%	98%		
Developer contribution to equity	40%	2%		
Developer operating margin			$20/kW	
Lease payment reserve			6 months	
Debt interest rate (PV)		7.0%		7.0%
Debt term		18 years		18 years
DSCR		1.3		1.3
Debt closing costs		$450,000		$450,000
Debt closing fee		2.75%		2.75%
Insurance (% of installed cost)	0.50%	0.50%	0.50%	0.50%
Analysis period	25 years	25 years	25 years	25 years

7. RESULTS

7.1. Reference Case

LCOE is highly dependent on cost and performance inputs, but financial structure and cost of capital can also have a significant impact. Figure 6 highlights the range of results by technology.[21] The difference in cost of energy between the technologies is more pronounced in the all-equity financial structures (represented with filled-in markers) in large part due to the cost-of-equity assumptions for PV versus CSP technologies (reference case equity returns for CSP were three percentage points higher than for PV).

Additionally, the PV system demonstrated a more consistent LCOE across financing structures than both CSP-Trough and CSP-Tower systems. A $0.03 variation in real LCOE was noted for PV across financing structures, while real LCOE results varied by $0.08 for CSP-Tower and $0.10 for CSP-Trough. Overall, the CSP technologies were found to be more sensitive to financing assumptions due to their capital intensity.

For all of the technologies examined, structures with project-level debt (represented with open markers) provide cost savings over their all-equity counterparts, despite the higher equity returns of 2%, or 200 basis points, required by tax investors when debt is introduced.

Table 4. Key Financing Variables by Financial Structure – CSP Reference Case

Financing Variable	All-Equity Partnership Flip	Leveraged Partnership Flip	Sale Leaseback	Single Owner
(Tax equity) IRR target year	9	9	20	20
(Tax equity) IRR target	12.0%	14.0%	12.0%	14.0%
Equity closing costs	$300,000	$300,000	$300,000	
Development fee	3%	3%	3%	
Tax investor contribution to equity	60%	98%		
Developer contribution to equity	40%	2%		
Developer operating margin			$20/kW	
Lease payment reserve			6 months	
Debt interest rate (CSP)		8.0%		8.0%
Debt term		18 years		18 years
DSCR		1.3		1.3
Debt closing costs		$450,000		$450,000
Debt closing fee		2.75%		2.75%
Insurance (% of installed cost)	0.50%	0.50%	0.50%	0.50%
Analysis period	25 years	25 years	25 years	25 years

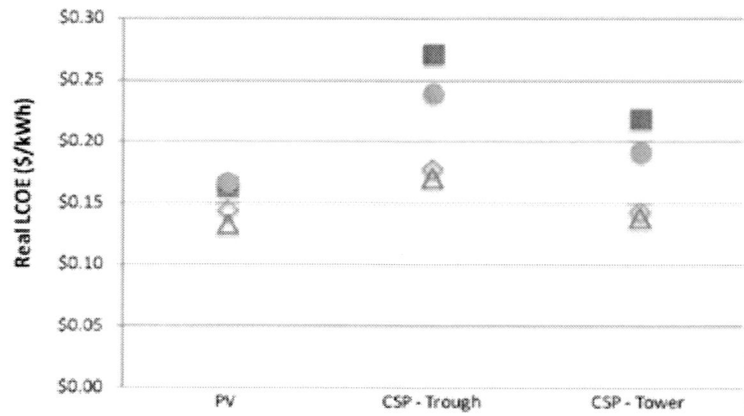

	PV	CSP - Trough	CSP - Tower
All Equity Flip	$0.163	$0.271	$0.220
Leveraged Flip	$0.145	$0.177	$0.144
Sale Leaseback	$0.168	$0.240	$0.193
Single Owner	$0.134	$0.171	$0.139

Figure 6. Financial structure impact on CSP and PV real LCOE (reference case).

CSP-Trough appears to be more expensive under most financial structures tested, but the results should be viewed cautiously as they are highly dependent on cost and other assumptions applied (certain sensitivities were assessed below). This analysis assumes the default values for cost and power-generation embedded in the SAM model. This analysis also provides no *energy value* premium for the CSP technologies associated with their energy storage and dispatchability, although recent analyses have shown time-of-day production scheduling can improve project economics (Denholm 2011).

For a generic PV plant, holding project-level debt reduced LCOE by $0.03–$0.07/kWh (20%– 50%) relative to the all-equity structures simply due to the availability of low-cost credit at a 7.0% interest rate. CSP-Trough and CSP-Tower structures with project-level debt—assuming a debt interest rate of 8%—provided an advantage in LCOE relative to all-equity structures of $0.06–$0.12/kWh (29% – 35%), even with significant debt closing costs of $450,000 and debt acquisition fees representing 2.75% of the debt principal.

Table 5 compares three commonly used metrics for the assessed PV project: real LCOE, nominal LCOE, and first-year PPA price. Real LCOE represents a level price over the analysis period, which produces the required IRRs that excludes the effect of inflation (i.e., represents current purchasing power). Nominal LCOE is similar to real LCOE but includes the effect of inflation. The first-year PPA price incorporates a 1% annual increase in the price-of-power produced. In all cases, all power is valued the same regardless of time-of-day or season it is produced. Similar comparisons of nominal LCOE, real LCOE, and first-year PPA are in the appendix. Only real LCOE is used in the remaining main body of the report.

7.2. Scenario and Sensitivity Analysis

Numerous single- and multi-variable scenario analyses were run to test the impact of alternative financial inputs and other key parameters on the LCOE. The following section describes scenarios examining the impacts of: (1) low interest rates such as those that could be obtained through a DOE loan guarantee, (2) changes in the IRR target year, and (3) increased cost of tax equity. Additionally, multi-variable low- and high-financing costs scenarios were developed.

7.2.1. Single Variable Scenarios

7.2.1.1. DOE Loan Guarantee Case

A separate scenario analysis was developed to assess the financing conditions like those under the DOE's loan guarantee program with a substantially lower interest rate for debt.

Twelve solar-electric generating projects (seven PV and five CSP) have received loan guarantees from DOE (DOE 2011). As a result, the cost of debt for these projects is expected to be substantially lower than market rates.[22] In the past, DOE has indicated its preference for simplistic ownership structures to avoid exacerbating an already-complex due diligence process (Wilson and Sonsini 2009). For this reason, we examined DOE loan guarantee scenarios utilizing only the single-owner financial structure.

For the loan guarantee case, debt returns (interest rates) were assumed to be 4.0%. The loan duration, or tenor, was held fixed for a period of 18 years. Debt closing costs were adjusted upwards from $450,000 to $1 million to reflect the potentially higher transaction costs of the loan guarantee program. These costs likely vary considerably by project and are only illustrative for the purpose of this analysis.

Table 5. Comparison of Real LCOE, Nominal LCOE, and First-Year PPA Price for a Theoretical PV Plant ($/kWh)

Metric	Single Owner	All-Equity Partnership Flip	Leveraged Partnership Flip	Sale Leaseback
Real LCOE (published)	$0.13	$0.16	$0.14	$0.17
Nominal LCOE	$0.16	$0.20	$0.18	$0.21
First Year PPA Price	$0.15	$0.19	$0.17	$0.19

The loan guarantee scenario led to $0.025/kWh–$0.03/kWh savings (17%) compared to the single-owner structure reference cases for the two CSP technologies, and a $0.02/kWh (15%) savings for PV (see Table 6).

7.2.1.2. Adjusted Target Year for IRR Return

Target year for the IRR return stands out as having a significant impact on the resulting LCOE, particularly for the all-equity partnership-flip structure. If we alter the target year by making it just one year shorter (or longer) for the all-equity partnership flip, the LCOE increases (or decreases) by nearly $0.06/kWh–$0.09/kWh (Table 7). This is because the all-equity structure needs the full-time horizon represented by the target year to meet the required tax equity return. Altering the target year significantly impacts the amount of cash that must be raised—via the PPA price—to meet the target return.

For the leveraged partnership-flip structure, the impact is insignificant. In this case, the availability of debt which has a lower yield requirement and longer payment time horizon frees up cash in early years to pay the tax equity investor. Altering the target year from the default value has no impact because the PPA price produced adequate cash flow to pay the negotiated return to the tax equity investor.

Table 6. LCOE Impact of Loan Guarantee Financing Case – PV, CSP-Trough, and CSP-Tower ($/kWh)

	Single-Owner Reference Case ($/kWh Real)	DOE Loan Guarantee ($/kWh Real)	Difference ($/kWh Real)	% Difference
PV	$0.134	$0.114	-$0.020	-15%
CSP-Trough	$0.171	$0.138	-$0.033	-19%
CSP-Tower	$0.139	$0.114	-$0.025	-18%

Table 7. LCOE Impact of Altered Target Year for PV ($/kWh Real)

Technology	Target Year Treatment	All-Equity Partnership Flip	Leveraged Partnership Flip
PV	**Ref. Case Target Year (year 9)**	19.9	16.0
	Target Year minus 1 (year 8)	25.4	16.0
	Target Year plus 1 (year 10)	16.8	16.0
CSP-Trough	**Ref. Case Target Year (year 9)**	34.0	22.6
	Target Year minus 1 (year 8)	39.6	22.6
	Target Year plus 1 (year 10)	28.6	22.6
CSP-Tower	**Ref. Case Target Year (year 9)**	27.3	18.2
	Target Year minus 1 (year 8)	31.4	18.2
	Target Year plus 1 (year 10)	23.1	18.2

Of course, it is unlikely that a tax investor will accept delayed receipt of an expected return without adjustment in other components of the deal structure. Instead, as general investment requirements hold, longer investment durations require higher yields to attract capital, all other things being equal. For example, 20-year bonds almost always provide a higher yield to the investor than bonds of 10- or 5-year maturities.

7.2.1.3. Increased Cost of Tax Equity

Several interviewees indicated their significant concern about the future supply and demand of tax equity and potential increases in the cost of tax equity after 2012. The federal 1603 Treasury grant program expires at the end of 2011, although projects completed after 2012 and before the applicable termination date (January 1, 2017, for solar) may qualify for the 1603 Treasury grant if those projects have "begun construction" in 2009, 2010, or 2011. As the 1603 Treasury grant ends, developers will need to find tax equity to finance their projects and monetize the ITCs. Due to the sharp increase in demand and an expected flat level in the supply of tax equity, the price of tax equity is expected to increase. Interviewees indicated that an increase of 2%–4% (200–400 basis points) is possible in the required yield. Tax equity returns can also be affected by other investment opportunities as well as the enactment of other policies.

Even at a 2%-yield increase, the LCOE impact is significant, particularly among the non-levered structures. For example, when 2% higher tax equity returns are required, the PV LCOE increases by $0.025/kWh for sale leaseback structures and $0.05/kWh for the all-equity partnership flip. In comparison, the change in LCOE for the structures that incorporate debt is modest at $0.003/kWh. If such an increase in equity costs were to happen, this could lead to greater interest in leveraged structures. However, levered structures may only be feasible for large utility-scale projects or a significant portfolio of projects that engages sufficient capital and holds PPAs with creditworthy entities.

The availability of both debt and equity capital may also be impacted by other policies (e.g., feed-in tariff, centralized clean energy financing bank, and carbon) and other competing tax equity investments (e.g., affordable housing) as well as non-tax-equity investments represented by the broader equities and commodities markets.

7.2.2. Low and High Financing Cost Scenarios

Two multi-variable alternate financing cost scenarios were developed to assess the combined impact of modifying several financing variables, including: tax equity return, IRR target year, debt interest rate, and DSCR (see Table 9). Modified inputs were adjusted consistently across the three technologies assessed even though they started from unique, technology-specific points. All other input parameters were held constant.

Figure 7 presents the range of results between the low and high scenarios across technologies and financial structures. Notably, low-cost scenarios produce LCOE results ranging from $0.12/kWh–$0.22/kWh, depending on the financial structure employed, indicating that the cost advantage of using project-level debt diminishes greatly as the cost of tax equity declines. In the high-cost case, the difference between the LCOE across financial structures is much wider, ranging from $0.014/kWh–$0.32/kWh. The all-equity structures result in the widest spread in LCOE between the low- and high-cases ($0.09/kWh–$0.14/kWh of difference).

Table 8. LCOE Impact of Increased Cost of Equity (PV only) ($/kWh)

	Single Owner	All-Equity Partnership Flip	Leveraged Partnership Flip	Sale Leaseback
Default Tax Equity Returns	$0.134	$0.163	$0.145	$0.168
Plus 2%	$0.137	$0.213	$0.148	$0.193
Difference ($/kWh)	$0.003	$0.050	$0.03	$0.025

Table 9. Modified Inputs for Low and High Scenarios (for all technologies)

Modified Inputs	Reference Case Value	Low Financing Cost Scenario	High Financing Cost Scenarios
Tax Equity Return	9%–11%, PV 12%–14%, CSP	-100 bp[a]	+100 bp
IRR Target Year	Yr. 9, unlevered Yr. 20, levered	+ 1 year	- 1 year
Debt interest rate	7%, PV 8%, CSP	- 100 bp	+ 100 bp
DSCR	1.3	- 0.1	+ 0.1

[a] bp = basis points, 100 basis points equal 1 percentage point.

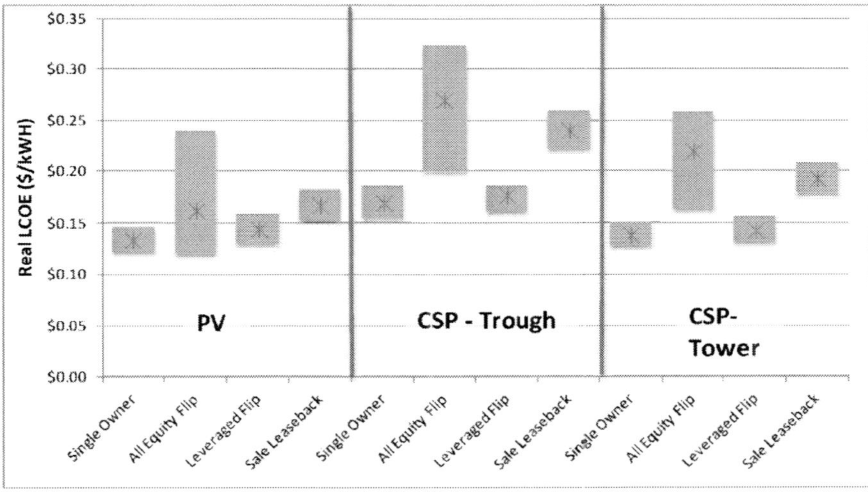

(* marker represents reference case value).

Figure 7. LCOE spreads between low and high cases for PV, CSP-Trough, and CSP-Tower.

The all-equity partnership-flip model yields the most expensive cost of energy in both the reference- and high-financing solar-cost cases. However, due to perceived risks associated with the technology, energy off-taker, transmission access, or other aspects of the project, project- level debt may simply not be available, available at a yield that eliminates its benefits, or available only with conditions that place extraordinary risk on the equity investors as to negate their interest. Accordingly, the relative benefits of a financial structure are very specific to the project and the parties involved as perceived risk and expected return are very unique attributes.

A similar trend can be seen when comparing the LCOE for CSP-Trough and CSP-Tower plants under the low- and high-financing cost scenarios. Across the board, structures with project-level debt yield the lowest LCOE and the narrowest spread between the low and high financing-cost cases. The high financing-cost case, which assumes a 14% target IRR (16% when project-level debt is included), significantly increases the LCOE for a CSP-Trough plant to more than $0.40/kWh.

The low-cost financing case highlights the potential for CSP projects to lower relevant power costs as risk perception—and thus required financing yields—declines in response to expected technology and developer experience. That is, as these technologies gain development and operational experience, the perceived risk—and required yield—is expected to decline. Lower financing costs, in response, should allow project developers deploying CSP technologies to offer lower LCOEs to their utility customers.

CONCLUSION

This analysis is intended to shed light on the direction and magnitude of the impact of solar project financial variables on LCOE when using four common renewable-financing structures. Differences in LCOE between PV, CSP-Trough, and CSP-Tower technologies are highly dependent on technology cost and performance assumptions as well as project-specific factors. Note that this study considered only the cost of energy generated, with no adjustments for time- of-delivery or power dispatchability.

Some results are not surprising because they are not unique to renewable projects or specific to these financing structures. Overall, this analysis found that all-equity structures yield higher LCOEs for CSP-Troughs, CSP-Towers, and PV than do financial structures that include project- level debt. Also, the all-equity structures are more sensitive to changes in key equity-related variables, such as expected return on equity and flip-target year. The LCOE for debt structures is less sensitive to changes in debt rates, as evidenced by the DOE loan guarantee case, where a substantial reduction in debt rate results in a relatively modest change in the overall LCOE (this ignores the other non-cost implications of the loan guarantee, such as potentially increased debt availability).

The choice of financing structure is influenced by a variety of factors that are project dependent and mostly influenced by the investing parties. These include risk tolerance, comfort with the financing structure, ability to use tax credits efficiently, project size, and projected output. Interviews with financial experts indicated that the selection of a financial structure is frequently based on non-cost considerations or project-specific risk parameters.

According to the analysis, financial structures employing project-level debt generally allow for a lower overall cost of capital and power costs. However, project-level debt can complicate the deal structure and is often available only to the largest, most secure developers. Although debt leverage can increase profitability of the project to the developer, equity investors view the presence of debt—and the lender's senior title to the assets—as a source of increased risk in the case of bankruptcy or underperformance of the project. Accordingly, tax equity investors require increased compensation through higher equity returns when debt is present. Further, the due diligence and negotiation processes can be more complex, potentially placing the project's success at greater risk. One interviewee in the PV industry indicated that it is difficult to negotiate debt with tax equity investors, but as deals get larger in the future, there could be more reason to bring in debt. One interviewee indicated that debt lenders prefer to be engaged only in larger projects of $25 million–$50 million or larger.

Interviewees indicated that all four of the financing structures have been used in the PV industry, although only limited data are available on the distribution of usage. In particular, sale leasebacks are common in the PV industry where, in some cases, it is leveraged and others not. For CSP projects, most current projects benefit from the federal loan guarantee program, so all-equity structures are not currently being utilized. The substantial reduction in borrowing rates from the loan guarantee program will also affect the debt-to-equity ratio and differ from a ratio based on market rates. However, scenarios based on market rates can have relevance for future projects once the loan guarantees are no longer available.

One key issue for the industry is potential increases in the cost of tax equity when the 1603 Treasury cash grant program expires. Interviewees suggest that tax equity investor returns will increase by perhaps 2%–4% (200–400 basis points) due to the increased demand for limited tax equity and expectations of shortage in the market. This analysis suggests that a significant change in tax equity rates can have a significant impact on the LCOE of projects using all-equity financing. Thus, potential increases in tax equity returns due to market shortages could lead to an increased interest in debt structures.

APPENDIX A. TABLES OF RESULTS

Table A-1. Reference Case Results ($/kWh)

	Primary Result	Single Owner	All- Equity Partnership Flip	Leveraged Partnership Flip	Sale Leaseback
PV	1st Year PPA Price	$0.153	$0.187	$0.166	$0.192
	Nominal LCOE	$0.164	$0.201	$0.178	$0.206
	Real LCOE	$0.134	$0.163	$0.145	$0.168
CSP-Trough	1st Year PPA Price	$0.167	$0.265	$0.174	$0.235
	Nominal LCOE	$0.212	$0.336	$0.219	$0.297
	Real LCOE	$0.171	$0.271	$0.177	$0.240
CSP-Tower	1st Year PPA Price	$0.136	$0.215	$0.141	$0.189
	Nominal LCOE	$0.172	$0.272	$0.178	$0.239
	Real LCOE	$0.139	$0.220	$0.144	$0.193

Table A-2. Low Case Results ($/kWh)

	Primary Result	Single Owner	All- Equity Partnership Flip	Leveraged Partnership Flip	Sale Leaseback
PV	**1st Year PPA Price**	$0.140	$0.139	$0.149	$0.176
	Nominal LCOE	$0.150	$0.149	$0.160	$0.189
	Real LCOE	$0.122	$0.121	$0.130	$0.153
CSP-Trough	**1st Year PPA Price**	$0.153	$0.196	$0.158	$0.217
	Nominal LCOE	$0.194	$0.248	$0.200	$0.275
	Real LCOE	$0.156	$0.201	$0.162	$0.222
CSP-Tower	**1st Year PPA Price**	$0.125	$0.161	$0.129	$0.175
	Nominal LCOE	$0.158	$0.203	$0.163	$0.221
	Real LCOE	$0.128	$0.164	$0.132	$0.179

Table A-3. High Case Results ($/kWh)

	Primary Result	Single Owner	All- Equity Partnership Flip	Leveraged Partnership Flip	Sale Leaseback
PV	**1st Year PPA Price**	$0.167	$0.275	$0.183	$0.209
	Nominal LCOE	$0.179	$0.295	$0.196	$0.225
	Real LCOE	$0.146	$0.240	$0.159	$0.183
CSP-Trough	**1st Year PPA Price**	$0.182	$0.317	$0.182	$0.254
	Nominal LCOE	$0.231	$0.401	$0.230	$0.321
	Real LCOE	$0.186	$0.324	$0.186	$0.259
CSP-Tower	**1st Year PPA Price**	$0.148	$0.254	$0.154	$0.204
	Nominal LCOE	$0.187	$0.320	$0.195	$0.257
	Real LCOE	$0.151	$0.259	$0.157	$0.208

REFERENCES

Bolinger, M. (2009). *Financing Non-Residential Photovoltaic Projects: Options and Implications*. LBNL-1410E. Berkeley, CA: Lawrence Berkeley National Laboratory. http://eetd.lbl.gov/ea/ems/reports/lbnl-1410e. pdf. Accessed March 19, 2012.

Cory, K. & Schwabe, P. (October 2009). *Wind Levelized Cost of Energy: A Comparison of Technical and Financing Input Variables*. NREL/TP-6A2-46671. Golden, CO: National Renewable Energy Laboratory. http://www. nrel.gov/docs/fy10osti/46671.pdf. Accessed March 19, 2012.

U.S. Department of Energy. "The Financing Force Behind America's Clean Energy Economy." Loan Program's Office website. https://lpo. energy.gov/?page_id=45. Accessed October 20, 2011.

Harper, J., Karcher, M. & Bolinger, M. (September 2007). *Wind Project Financing Structures: A Review & Comparative Analysis*. LBNL-63434. Berkeley, CA: Lawrence Berkeley National Laboratory. http://eetd.lbl. gov/ea/emp/reports/63434.pdf. Accessed March 19, 2012.

Kolb, G. J., Ho, C. K., Mancini, T. R. & Gary, J .A. (2011). *Power Tower Technology Roadmap and Cost Reduction Plan*. SAND2011-2419. Albuquerque, NM: Sandia National Laboratory.

Scharfenberger, P. (2010). "*Normal Accounting Rules Limit Utility Ownership of Renewable Energy Projects*." https://financere.nrel.gov/ finance/ content/normal-accounting-rules-limit-utility-ownership-renewable-energy-projects. Accessed November 21, 2011.

Turchi, C., Mehos, M.;, Ho, C. K. & Kolb, G. J. (2010). *Current and Future Costs for Parabolic Trough and Power Tower Systems in the US Market.* NREL/CP-5500-49303. Golden, CO: National Renewable Energy Laboratory. http://www.nrel.gov/docs/fy11osti/49303.pdf. Accessed March 19, 2012.

Wilson Sonsini. (2009). "FIPP, FIPP, Hooray ... ? Analysis of and Commentary on DOE's Financial Institution Partnership Program, October 28, 2009." http://www.wsgr.com/WSGR/Display.aspx?SectionName= publications/PDFSearch/wsgralert_FI PP.htm. Accessed November 21, 2011.

Zhang, Y. & Smith, S. J. (2008). *Long-Term Modeling of Solar Energy: Analysis of Concentrating Solar Power (CSP) and PV Technologies.* PNNL-16727. Richland, WA: Pacific Northwest National Laboratory.

End Notes

[1] Lease structures have been used by the solar industry under earlier investment tax credits.

[2] See Section 1603 of Division B of the American Recovery and Reinvestment Act of 2009, as amended.

[3] The federal ITC enables developers to obtain a 30% tax credit for expenditures associated with installing qualifying renewable energy facilities installed by the end of 2016. Facilities qualifying for the 30% ITC include solar, small wind, and fuel cells. The PTC is a federal tax credit of $0.022/kilowatt-hour (kWh) inflation adjusted for the first 10 years of generation from qualifying renewable energy facilities. The PTC is available for wind project through the end of 2012 and for other qualifying technologies such as biomass, geothermal, and marine and hydrokinetic resources through 2013. The PTC is not applicable to solar projects. The MACRS depreciation incentive allows projects to depreciate projects over an accelerated five-year schedule. A 50% first year bonus depreciation provision is currently available through the end of 2012. For more information on these tax incentives, see the Database of State Incentives for Renewable Energy (DSIRE) at http://dsireusa.org/incentives/incentive.cfm?Incentive_Code=US06F&re=1&ee=1. Accessed December 13, 2011.

[4] Because the cash grants are scheduled to expire, this analysis focuses on incentives that require tax equity investment to take full advantage of the tax incentives.

[5] SAM is available at https://www.nrel.gov/analysis/sam/. Accessed September 26, 2011.

[6] This table does not represent all of the projects using financial structure in today's market. Actual market projects may have very unique features.

[7] Shaded areas represent disbursement of benefits and assumed allocations among investors.

[8] Note that a single owner is usually not a utility due to Internal Revenue Service normalization rules (see Scharfenberger 2011).

[9] All illustrated percentage values in the text and financial structure graphics are based on reported cases. Actual percentages relevant to a specific project may be very different based on the investors and their relevant risk tolerances and reward expectations.

[10] Values to the left of the slash indicate the allocations before the flip point. Values to the right of the slash indicate allocations after the flip point.

[11] The IRS guidelines specifically apply to wind facilities, but various entities in the solar industry believe the guideline principles are applicable to solar projects as well. See Rev. Proc. 2007-65.

[12] The SAM model allows the user to select the development fee to be recovered over a set period of time or immediately upon project commercial operation.

[13] The SAM model employed for this analysis is only capable of examining flip structures with a single-flip point.

[14] Because of their capital-intensive nature, renewable energy projects are typically financed using non-recourse, or project, financing under which a special purpose entity is created to shield the sponsor from the detrimental effects of project failure. Under non-recourse financing, the project assets are paid for entirely from project cash flow, not the sponsor's balance sheet. In contrast, under a corporate finance structure, loans are secured from the general assets or creditworthiness of the project sponsors.

[15] There is no specific method to assess joint ventures in the SAM model. Instead, one could utilize the single- owner structure and input the weighted average cost of equity among the joint venture parties as the required return on equity.

[16] The analysis was conducted using SAM Version 2011.6.30.

[17] The SAM User Guide (Gilman 2011) provides more information about the model and its historical development.

[18] The model is not designed to provide a bankable pro forma. The pro forma is an annual model, as opposed to a quarterly or monthly model generally required of banking institutions.

[19] Structures with debt carry a greater default risk for the tax equity investor because the project's debt lenders get priority rights in the case of project default. Therefore, the tax equity investor in the debt-leveraged structures requires a higher target IRR relative to the all-equity financing structure, which compensates for their increased risk exposure (Cory and Schwabe 2009).

[20] DSCRs were held constant across CSP and PV technologies but interest rates were adjusted to account for technology-related risk.

[21] Note that the LCOE values for PV and CSP are also affected by the size of the hypothetical project considered (20 MW for PV and 100 MW for CSP); the fixed costs associated with the smaller PV project increases the LCOE on a relative basis.

[22] For rates associated with the loans made by the Federal Finance Bank, see http://www.treasury.gov/ffb/press_releases/2011/07-2011.shtml.

INDEX

A

B

C

D

E

F

G

H

I

J

L

M

N

O

P

Q

R

S

T